S0-BCN-128

GEOMETRIC TRANSFORMATIONS II

NEW MATHEMATICAL LIBRARY

published by
Random House and The L. W. Singer Company
for the Monograph Project *of the*
SCHOOL MATHEMATICS STUDY GROUP†

EDITORIAL PANEL

Ivan Niven, Chairman (1964–1968)
University of Oregon

Anneli Lax, Technical Editor
New York University

E. G. Begle	*Stanford University*
M. Bell (1962–68)	*University of Chicago*
L. Bers (1958–62)	*Columbia University*
B. Bold (1963–66)	*Stuyvesant High School, N. Y.*
W. G. Chinn (1961–67)	*San Francisco Public Schools*
H. S. M. Coxeter (1958–61)	*University of Toronto*
P. J. Davis (1961–64)	*Brown University*
E. C. Douglas (1966–69)	*The Taft School, Conn.*
E. Dyer (1963–66)	*City University of New York*
B. Gordon (1966–69)	*University of California, Los Angeles*
H. J. Greenberg (1964–67)	*University of Denver*
P. R. Halmos (1958–63)	*University of Hawaii*
J. H. Hlavaty (1958–63)	*DeWitt Clinton High School, N. Y.*
N. Jacobson (1958–61)	*Yale University*
M. Kac (1961–65)	*Rockefeller University*
M. Klamkin (1965–68)	*The Ford Motor Company*
J. Landin (1966–69)	*University of Illinois, Chicago*
R. S. Pieters (1958–62)	*Phillips Academy*
H. O. Pollak (1958–61)	*Bell Telephone Laboratories, Inc.*
G. Pólya (1958–61)	*Stanford University*
W. Prenowitz (1962–65)	*Brooklyn College*
D. Richmond (1965–68)	*Williams College*
A. Y. Rickey (1965–68)	*Dade County, Fla., Public Schools*
H. E. Robbins (1958–61)	*Columbia University*
W. W. Sawyer (1958–60)	*University of Toronto*
D. S. Scott (1963–66)	*Stanford University*
N. E. Steenrod (1958–62)	*Princeton University*
J. J. Stoker (1958–61)	*New York University*
M. Zelinka (1962–65)	*Weston High School, Mass.*
L. Zippin (1958–61)	*City University of New York*

Arlys Stritzel, Editorial Assistant

† The School Mathematics Study Group represents all parts of the mathematical profession and all parts of the country. Its activities are aimed at the improvement of teaching of mathematics in our schools. Further information can be obtained from: School Mathematics Study Group, Cedar Hall, Stanford University, Stanford, California 94305.

GEOMETRIC TRANSFORMATIONS II

by

I. M. Yaglom

translated from the Russian by

Allen Shields
University of Michigan

21

RANDOM HOUSE

THE L. W. SINGER COMPANY

113501

Illustrated by George H. Buehler

First Printing

Library of Congress Catalog Card Number: 67-20607

Manufactured in the United States of America

Note to the Reader

This book is one of a series written by professional mathematicians in order to make some important mathematical ideas interesting and understandable to a large audience of high school students and laymen. Most of the volumes in the *New Mathematical Library* cover topics not usually included in the high school curriculum; they vary in difficulty, and, even within a single book, some parts require a greater degree of concentration than others. Thus, while the reader needs little technical knowledge to understand most of these books, he will have to make an intellectual effort.

If the reader has so far encountered mathematics only in classroom work, he should keep in mind that a book on mathematics cannot be read quickly. Nor must he expect to understand all parts of the book on first reading. He should feel free to skip complicated parts and return to them later; often an argument will be clarified by a subsequent remark. On the other hand, sections containing thoroughly familiar material may be read very quickly.

The best way to learn mathematics is to *do* mathematics, and each book includes problems, some of which may require considerable thought. The reader is urged to acquire the habit of reading with paper and pencil in hand; in this way mathematics will become increasingly meaningful to him.

For the authors and editors this is a new venture. They wish to acknowledge the generous help given them by the many high school teachers and students who assisted in the preparation of these monographs. The editors are interested in reactions to the books in this series and hope that readers will write to: Editorial Committee of the NML series, NEW YORK UNIVERSITY, THE COURANT INSTITUTE OF MATHEMATICAL SCIENCES, 251 Mercer Street, New York, N. Y. 10012.

The Editors

v

NEW MATHEMATICAL LIBRARY

Other titles will be announced when ready

1. NUMBERS: RATIONAL AND IRRATIONAL by Ivan Niven
2. WHAT IS CALCULUS ABOUT? by W. W. Sawyer
3. INTRODUCTION TO INEQUALITIES by E. Beckenbach and R. Bellman
4. GEOMETRIC INEQUALITIES by N. D. Kazarinoff
5. THE CONTEST PROBLEM BOOK I, Annual High School Contests of the Mathematical Association of America, 1950–1960, compiled and with solutions by Charles T. Salkind
6. THE LORE OF LARGE NUMBERS by P. J. Davis
7. USES OF INFINITY by Leo Zippin
8. GEOMETRIC TRANSFORMATIONS I by I. M. Yaglom, translated from the Russian by Allen Shields
9. CONTINUED FRACTIONS by C. D. Olds
10. GRAPHS AND THEIR USES by Oystein Ore
11. HUNGARIAN PROBLEM BOOK I, based on the Eötvös Competitions, 1894–1905
12. HUNGARIAN PROBLEM BOOK II, based on the Eötvös Competitions, 1906–1928
13. EPISODES FROM THE EARLY HISTORY OF MATHEMATICS by Asger Aaboe
14. GROUPS AND THEIR GRAPHS by I. Grossman and W. Magnus
15. MATHEMATICS OF CHOICE by Ivan Niven
16. FROM PYTHAGORAS TO EINSTEIN by K. O. Friedrichs
17. THE MAA PROBLEM BOOK II, Annual High School Contests of the Mathematical Association of America, 1961–1965, compiled and with solutions by Charles T. Salkind
18. FIRST CONCEPTS OF TOPOLOGY by W. G. Chinn and N. E. Steenrod
19. GEOMETRY REVISITED by H. S. M. Coxeter and S. L. Greitzer
20. INVITATION TO NUMBER THEORY by O. Ore
21. GEOMETRIC TRANSFORMATIONS II by I. M. Yaglom, translated from the Russian by Allen Shields

Contents

Translator's Preface 1

From the Author's Preface 2

Introduction. **What is Geometry?** 4

Chapter I. **Classification of Similarity Transformations** 9

1. Central similarity (homothety) 9

2. Spiral similarity and dilative reflection. Directly similar and oppositely similar figures 36

Chapter II. **Further Applications of Isometries and Similarities** 63

1. Systems of mutually similar figures 63

2. Applications of isometries and of similarities to the solution of maximum-minimum problems 84

Solutions. Chapter One. Classification of similarities 87

Chapter Two. Further applications of isometries and similarities 141

Comparison of problem numbers between the Russian edition (1955)
and the English translation

Problems that have been added for the English edition have the word "new" next to them.

Russian	English	Russian	English	Russian	English
44	1	59	29	81	57
45	2	60	30	82	58
new	3	61	31	83	59
new	4	62	32	84	60
new	5	new	33	85	61
new	6	63	34	86	62
new	7	64	35	87	63
46	8	65	36	88	64
47	9	66	37	89	65
48	10	new	38	90	66
new	11	new	39	91	67
new	12	new	40	92	68
49	13	67	41	93	69
50	14	68	42	94	70
new	15	69	43	95	71
new	16	70*	44*	96	72
51	17	new	45	97	73
52	18	new	46	98	74
53	19	71	47	99	75
54	20	72	48	100	76
55	21	73	49	101	77
56	22	74	50	102	78
new	23	75	51	103	79
new	24	76	52	104	80
new	25	77	53	105	81
new	26	78	54	new	82
57	27	79	55	106	83
58	28	80	56		

* Problem 70 in the Russian edition has been redone and a new part has been added to form Problem 44 in this translation.

Translator's Preface

The present volume is Part II of *Geometric Transformations* by I. M. Yaglom. The English translation of Part I has already appeared in this series. In the original Russian edition (1955) the two parts were published together in a single volume, and Part III appeared as a separate volume. An English translation of Part III will appear presently.

This book is not a text in plane geometry; on the contrary, the author assumes that the reader already has some familiarity with the subject. Part II deals with similar figures and with transformations that preserve similarity.

As in Part I, the problems are the heart of the book. There are eighty-three problems in all, and the reader should attempt to solve them for himself before turning to the solutions in the second half of the book. The numbering of the problems is not the same in the English edition as in the original Russian. In the Russian the problems were numbered consecutively from 1 to 106 (Problems 1–47 were in Part I, and 48–106 were in Part II). In the English translation the numbering has started over again in Part II. A chart comparing the numbering in the original and the translation has been included, since in Part I there are several references to problems in Part II, and these references used the problem numbers from the Russian original.

Those footnotes preceded by the usual symbols † or ‡ were taken over from the Russian edition (or were added by the author for the American edition), while those preceded by the symbol T have been added by the translator.

The translator wishes to thank Professor Yaglom for his valuable assistance in preparing the American edition of his book. He read the manuscript of the translation and made a number of additions and corrections. Also, he prepared nineteen additional problems for the translation (these are indicated in the chart comparing the numbering of the problems).

The translator calls the reader's attention to footnote T on p. 13, which explains an unorthodox use of terminology in this book.

In conclusion the translator wishes to thank the members of the SMSG Monograph Project for the advice and assistance. Professor H. S. M. Coxeter was particularly helpful with the terminology. Especial thanks are due to Dr. Anneli Lax, the technical editor of the project, for her invaluable assistance, her patience, and her tact, and to her assistant, Arlys Stritzel.

Allen Shields

1

From the Author's Preface

This work, consisting of three parts, is devoted to elementary geometry. A vast amount of material has been accumulated in elementary geometry, especially in the nineteenth century. Many beautiful and unexpected theorems were proved about circles, triangles, polygons, etc. Within elementary geometry whole separate "sciences" arose, such as the geometry of the triangle or the geometry of the tetrahedron, having their own, extensive, subject matter, their own problems, and their own methods of solving these problems.

The task of the present work is not to acquaint the reader with a series of theorems that are new to him. It seems to us that what has been said above does not, by itself, justify the appearance of a special monograph devoted to elementary geometry, because most of the theorems of elementary geometry that go beyond the limits of a high school course are merely curiosities that have no special use and lie outside the mainstream of mathematical development. However, in addition to concrete theorems, elementary geometry contains two important general ideas that form the basis of all further development in geometry, and whose importance extends far beyond these broad limits. We have in mind the deductive method and the axiomatic foundation of geometry on the one hand, and geometric transformations and the group-theoretic foundation of geometry on the other. These ideas have been very fruitful; the development of each leads to non-Euclidean geometry. The description of one of these ideas, the idea of the group-theoretic foundation of geometry, is the basic task of this work . . . .

Let us say a few more words about the character of the book. It is intended for a fairly wide class of readers; in such cases it is always necessary to sacrifice the interests of some readers for those of others. The author has sacrificed the interests of the well prepared reader, and has striven for simplicity and clearness rather than for rigor and for logical exactness. Thus, for example, in this book we do not define the general concept of a geometric transformation, since defining terms that are intuitively clear always causes difficulties for inexperienced readers. For the same reason it was necessary to refrain from using directed angles and to postpone to the second chapter the introduction of directed segments, in spite of the disadvantage that certain arguments in the basic text and in the solutions of the problems must, strictly speaking, be considered incomplete . . . . It seemed to us that in all these cases the well prepared reader could complete the reasoning for himself, and that the lack of rigor would not disturb the less well prepared reader . . . .

The same considerations played a considerable role in the choice of terminology. The author became convinced from his own experience as a student that the presence of a large number of unfamiliar terms greatly increases the difficulty of a book, and therefore he has attempted to practice the greatest economy in this respect. In certain cases this has led him to avoid certain terms that would have been convenient, thus sacrificing the interests of the well prepared reader . . . .

The problems provide an opportunity for the reader to see how well he has mastered the theoretical material. He need not solve all the problems in order, but is urged to solve at least one (preferably several) from each group of problems; the book is constructed so that, by proceeding in this manner, the reader will not lose any essential part of the content. After solving (or trying to solve) a problem, he should study the solution given in the back of the book.

The formulation of the problems is not, as a rule, connected with the text of the book; the solutions, on the other hand, use the basic material and apply the transformations to elementary geometry. Special attention is paid to methods rather than to results; thus a particular exercise may appear in several places because the comparison of different methods of solving a problem is always instructive.

There are many problems in construction. In solving these we are not interested in the "simplest" (in some sense) construction—instead the author takes the point of view that these problems present mainly a logical interest and does not concern himself with actually carrying out the construction.

No mention is made of three-dimensional propositions; this restriction does not seriously affect the main ideas of the book. While a section of problems in solid geometry might have added interest, the problems in this book are illustrative and not at all an end in themselves.

The manuscript of the book was prepared by the author at the Orekhovo-Zuevo Pedagogical Institute . . . in connection with the author's work in the geometry section of the seminar in secondary school mathematics at Moscow State University.

I. M. Yaglom

INTRODUCTION

What is Geometry?

In the introduction to the first volume we defined geometry as the study of those properties of figures that are not changed by motions; motions were defined as transformations that do not change the distance between any two points of the figure (see pp. 10–11 of Volume One). It follows at once from this that the most important geometric properties of a figure seem to be the distances between its various points, since the concept of distance between points—the length of a segment—appears to be the most important concept in all geometry. However, if we examine carefully all the theorems of elementary geometry as presented in Kiselyov's text,[T] then we see that the concept of distance between points hardly figures at all in these theorems. All the theorems on parallel and perpendicular lines (for example, the theorems: "if two parallel lines are cut by a third line, then the corresponding angles are equal" or "from each point not on a given line there is one and only one perpendicular to the given line"), most of the theorems about circles (for example, "through three points not all lying on a straight line one and only one circle can be passed"), many of the theorems about triangles and polygons (for example, "the sum of the angles of a triangle equals a straight angle", or "the diagonals of a rhombus are perpendicular to each other and bisect the angles of the rhombus") have nothing whatsoever to do with the concept of distance. And even in those theorems whose formulation does contain the concept of length of a segment (for example, "the

[T] This is the standard textbook of plane geometry in the Soviet Union.

4

bisector of any angle of a triangle divides the opposite side into parts proportional to the adjacent sides of the triangle", "in a given circle or in congruent circles the longer of two unequal chords lies closer to the center", or even the Pythagorean theorem: "if the sides of a right triangle are measured in the same units, then the square of the length of the hypotenuse equals the sum of the squares of the lengths of the legs"), in actual fact it is not the length of some segment or other that plays a role, but only *the ratio of the lengths* of two or of several segments. It is easy to convince oneself of this if one thinks about the content of these theorems. For example, in the Pythagorean theorem it is not the actual lengths of the sides of the triangle that play a role, but only the ratios of the lengths of the legs to the length of the hypotenuse: this theorem says that if a right triangle ABC has legs of lengths b and c, and if k and l denote the ratios of these lengths to that of the hypotenuse a (that is, $b/a = k$, $c/a = l$), then $k^2 + l^2 = 1$.

It is not difficult to understand the general principle behind this. The concept of length of a segment uses in an essential manner the presence of some fixed unit of measurement for lengths; the numbers that express the lengths of one and the same segment will be different if the segment is measured in centimeters, or in kilometers, or in inches. But the content of geometric theorems cannot depend on the particular unit of measurement that has been chosen. It follows that in geometric theorems lengths cannot figure by themselves; instead we can encounter only the ratios of the lengths of two or of several segments (these ratios do not depend on the choice of the unit of measurement). Thus the previous formulation of the Pythagorean theorem beginning with the words: "if the sides of a right triangle are measured in the same units, then . . ." tells us that the theorem speaks of the ratios of the lengths of the sides of the triangle. If we know that the lengths of several segments are measured in the same units, but we do not know what the unit of measurement actually is, then we can consider only the *ratios* of the lengths of these segments. It would, of course, be useless to require in the hypothesis of a theorem that a segment be measured in a definite unit of measurement, for example in meters; clearly a theorem cannot be true only when the segment is measured in centimeters and false when it is measured in, say, inches.

This is connected with the fact that, from the geometric point of view, all segments are the same, no one of them is in any way distinguished or preferred; therefore all definitions of a unit of length are of a completely arbitrary character from the geometric point of view. For example, the *meter* is defined as the length of a certain platinum-iridium bar, kept in the Bureau of Weights and Measures in Paris; it is also defined as 1,552,734.83 times the wave length of the red line of emission of cadmium under certain standard conditions. As another example, the English *yard* was originally introduced as the distance from the nose of the English King Henry the First to the tip of the middle finger of his outstretched hand. It is natural, therefore, that in the formulation of geometric theorems one does not use the lengths of segments, but merely the ratios of lengths—quantities that do not depend on the choice of the unit of measurement.†

Thus the concept of distance between points, which according to our definition of geometry should play a basic role, actually does not appear directly in geometric theorems. This circumstance was already pointed out by F. Klein, the first to give a precise definition of geometry. Indeed, Klein's definition is somewhat different from that given in the introduction to Volume One. Here it is: *geometry is the science that studies those properties of geometric figures that are not changed by similarity transformations*. Similarity transformations can be defined as those *transformations that do not change the ratios of the distances between pairs of points;* this abstract definition of similarity transformation can be replaced by a complete description of all such transformations. Such a description will be given in Chapter I, Section 2 of this book. Klein's definition says that, in a definite sense, not only does geometry not distinguish between congruent figures, but it does not even distinguish between similar figures; indeed, in order to assert that two triangles are congruent, and not merely similar, we must have fixed a definite unit of measurement, once and for all, with which to measure the sides of both triangles. It is this very "indistinguishability" of similar figures that enables us to represent figures of large dimensions in a picture; the teacher uses this principle when he tells the students to reproduce "accurately" in their notebooks the figure he is drawing on the blackboard and which, of course, could not possibly fit into their notebooks without being reduced in size.‡

We see, then, that the basic role in elementary geometry is in fact played by similarity transformations; thus a special study of these transformations is important. Not only is the study of similarity transformations of great theoretical interest, it is also very useful in the solution of many different problems; indeed, in this respect similarity transformations do not take second place to isometries. As an example one may take the problem of constructing a quadrilateral, similar to a given one, whose points pass through four given points [see problem 55 (b) of Section 1, Chapter II]. This problem is a generalization of the following

---

† Note that in contrast to the lengths of segments, the magnitudes of angles often figure in the formulation of geometric theorems. This is because the unit of angular measure can be defined purely geometrically: the *radian* is defined as the central angle subtending an arc of a circle whose length is equal to that of the radius of the circle, and a *right angle* is defined to be an angle equal to its supplement. The difference between the concepts of length of a segment and magnitude of an angle is well illustrated, for example, by the following theorem: *in a right triangle, one of whose acute angles is* 30°, *the ratio of the length of the shortest side to the hypotenuse is* 1:2.

‡ Incidentally, in certain situations the definition of geometry given in Volume One is preferable to the one given here. This question is treated more fully in the introduction to Volume Three.

three familiar problems usually solved by using special properties of the square, rectangle and rhombus:
   (a) construct a square whose sides pass through four given points;
   (b) construct a rectangle with given ratio of sides, whose sides pass through four given points;
   (c) construct a rhombus with given angles, whose sides pass through four given points.

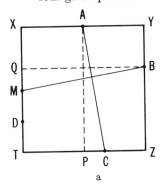 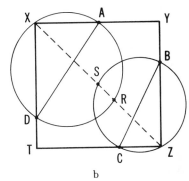

a   b

Figure 1

The first of these problems can be solved as follows: if the points A, B and C lie on the sides XY, YZ and ZT of the square $XYZT$ (Fig. 1a), and if the line through B perpendicular to AC meets the line XT in a point M, then $BM = AC$, since triangles ACP and BMQ in Figure 1a are congruent. Thus, if we know four points A, B, C and D lying on the four sides of the square, then we can find one more point M on side TX (or on its extension) and draw the line TX passing through the two points D and M (assuming $D \neq M$).

This same problem can be solved in another way. As before let A, B, C, D be points on the sides XY, YZ, ZT, TX of the square. The circle with AD as diameter passes through the point X; let R be the other point of intersection of the diagonal XZ with this circle. Likewise, the circle with diameter BC passes through Z; let S be the other point of intersection of the diagonal XZ with this circle (Figure 1b). Since XZ bisects the angle at X the point R is the midpoint of the circular arc ARD; likewise S is the midpoint of arc BSC. Thus the line XZ can be constructed as the line through the two midpoints R and S (if $R \neq S$). The other points of intersection of this line with the two circles will be the desired vertices X and Z.

The first solution outlined above can be generalized in a natural way to solve the second problem, (b), where now $AC:BM = AP:BQ = TX:XY$ are known ratios (see Figure 2a). The second solution to the problem of constructing the square can be generalized very nicely to solve the third problem, (c); in this case however, the circular arcs must be constructed on the segments AD and BC so that these segments subtend an inscribed angle equal to the given angle of the rhombus. The diagonal XZ is then found as before as the line through the midpoints R and S of these circular arcs (Figure 2b).

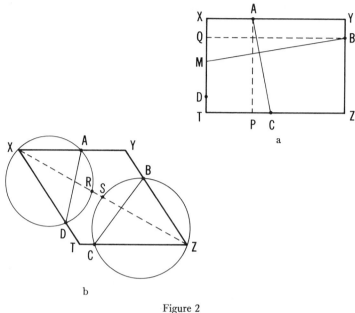

Figure 2

Although these solutions to the three problems (a), (b), (c) are very beautiful, they are somewhat artificial, and it is not so easy to hit upon such proofs for oneself. General considerations based on similarity transformations enable us to find a more natural solution to a more general problem [55(b)] which includes these three problems as special cases.

The reader will find many other problems that can be solved by the application of similarity transformations.

Note also that, since isometries are a special case of similarity transformations, many problems whose solutions use isometries can be greatly generalized if one solves them by similarity transformations instead. For example, the problem of constructing a polygon, given the vertices of isosceles triangles having given vertex angles and constructed on the sides of the polygon (we spoke of this problem in the introduction to Volume One), can be so generalized in the following manner:

In the plane n points are given; they are the outer vertices of triangles constructed on the sides of a certain n-gon (with these sides as bases) and similar to n given triangles. Construct the n-gon (see problem 37 of Section 2, Chapter I).

The reader will find many other such examples in Chapter I.

CHAPTER ONE

Classification of Similarity Transformations

1. Central similarity (homothety)

We say that a point A' is obtained from a point A by a *central similarity* (or by a *homothety*) with center O and similarity coefficient k if A' lies on the line OA, on the same side of the point O as A, and if $OA'/OA = k$ (Figure 3a). The transformation of the plane that carries each point A into the point A' centrally similar to A with respect to the center of similarity O and with similarity coefficient k is called a *central similarity* (or *homothety*, or *dilatation*); the point O is called the *center* and the number k is called the *coefficient* of the transformation. The point A' is called the *image* of A *under the transformation*. The images of all points of a figure F form a figure F', said to be centrally similar to F (with respect to the center of similarity O and with similarity coefficient k) (Figure 3b). Clearly the figure F in turn is centrally similar to F' (with respect to the same center of similarity and with coefficient $1/k$); this enables

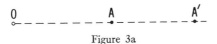

Figure 3a

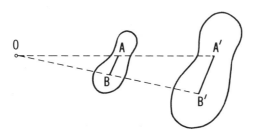

Figure 3b

9

one to speak of pairs of centrally similar figures. One says also that the figures F and F' are similar,† or that they are similarly placed.

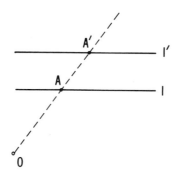

Figure 4a

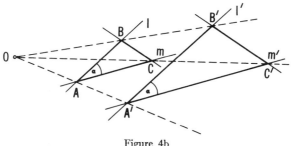

Figure 4b

A central similarity transformation carries a line l into a line l' parallel to l; to construct l' it is sufficient to find the image A', centrally similar to some point A on the line l, and to pass through A' a line parallel to l (Figure 4a). If two lines l and m intersect in an angle α, then their images l' and m' intersect in the same angle α; thus the triangle $A'B'C'$, centrally similar to a given triangle ABC has the same angles as ABC, that is, the two triangles are similar (Figure 4b). A circle S with center A and radius r is carried by a central similarity onto a new circle S' with center at the point A' into which the point A is taken, and with radius $r' = kr$, where k is the similarity coefficient (Figure 5). Indeed, from the similarity of triangles OAM and $OA'M'$ (where M is any point of S and M' its image) it follows that $A'M'/AM = k$, that is, $A'M' = kr$; this shows that S is carried onto the circle with center A' and radius kr.

---

† The transformation of central similarity is a similarity transformation in the sense of the definition given in the introduction, since in this transformation the lengths of all segments are multiplied by one and the same number k (see below, page 26).

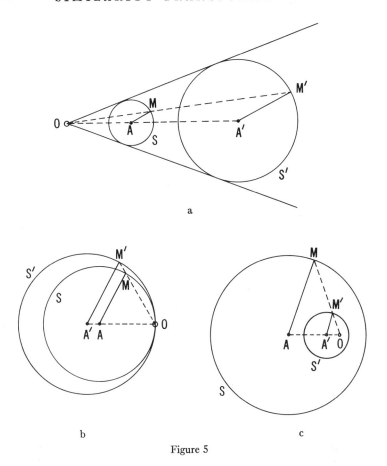

Figure 5

Clearly any two incongruent circles S and S' of radii r and r' and with centers at the points A and A' can be considered as centrally similar; it is sufficient to take for the center of similarity the point O lying on the line AA' outside of the segment AA', and such that $OA'/OA = r'/r$, and to take for the similarity coefficient the ratio r'/r (Figure 5). The point O is called the *exterior center of similarity* of the circles S and S' (as opposed to the interior center of similarity, of which we shall speak below). To construct the exterior center of similarity O of two incongruent circles S and S', it is sufficient to draw any two parallel and similarly directed radii AM and $A'M'$ in these circles and then to join M and M'; O is the point of intersection of AA' and MM'. If the smaller of the two circles does not lie inside the larger, then the exterior similarity center can also be found as the point of intersection of the common exterior tangents (Figure 5a); if S and S' are internally tangent, then the similarity center coincides with the point of tangency (Figure 5b).

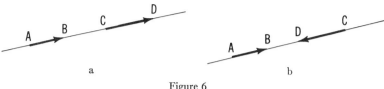

a   b

Figure 6

It is often convenient to consider that the ratio of two segments AB and CD, lying on one and the same line, has a definite sign: *the ratio AB/CD is said to be positive in case the directions of the segments AB and CD* (that is, the direction from the point A to the point B, and from C to D) *coincide* (Figure 6a), *and negative when these directions are opposite* (Figure 6b). It is clear that the order in which we write the endpoints of the interval is essential; thus, for example, $BA/CD = -AB/CD$. This convention about the signs of intervals is convenient in many geometric questions; we shall use it in the future.† If we accept this convention, then the similarity coefficient of two centrally similar figures can be either positive or negative. Namely, two figures F and F' will be said to be centrally similar with center of similarity O and (negative!) similarity coefficient $-k$, if each two corresponding points A and A' of these figures lie on a line through O and on opposite sides of O, and if the ratio of the lengths of the segments OA' and OA is equal to k (Figure 7); this condition can be written in the form $OA'/OA = -k$. A central similarity with similarity center O and negative similarity coefficient $-k$ is the same as the transformation obtained by first performing a central similarity with center O and positive coefficient k (carrying the figure F into F_1; see Figure 7), and then following this by a half turn about O (taking F_1 into F'); or the half turn about O can be performed first (taking the figure F into F_2; see Figure 7) followed by a central similarity with center O and positive coefficient k (taking F_2 into F').

† This definition of the sign of the ratio of intervals can be explained in the following manner. Choose on a line some direction as the positive direction (it can be indicated by an arrow placed on the line); an interval AB on this line will be considered positive if its direction (from the point A to the point B) is positive, otherwise it will be considered negative (compare the fine print at the bottom of page 20 and the top of page 21 in Volume One). Thus the ratio of two segments can be either positive or negative and, as one sees easily, this is independent of the particular direction on the line that was chosen as the positive direction; the ratio AB/CD will be positive if the segments AB and CD have the same sign (that is, either both intervals are positive or both are negative), and the ratio will be negative if the segments have opposite signs (that is, one of the segments is positive and the other is negative).

In the language of vectors our definition of the ratio of segments, taken with a definite sign, can be formulated as follows: $AB/CD = k$, where k is the positive or negative number such that $AB = kCD$.

Thus *a central similarity with negative similarity coefficient* $-k$ *is the sum*[T] *of a central similarity with the same center O and with positive coefficient k and a half turn about O*, taken in either order. In the future when we speak of central similarities we always understand that the similarity coefficient can be either positive or negative.

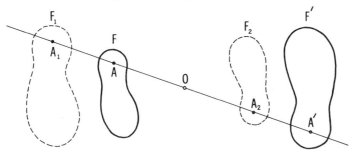

Figure 7

A central similarity with negative similarity coefficient $-k$ also carries a line l into a parallel line l' (but now the center of similarity lies between the two lines l and l'; see Figure 8), and it carries a circle S into another circle S' (the center A' of the circle S' is centrally similar to the center A of the circle S with negative similarity coefficient $-k$, and the ratio of the radii r'/r is equal to k; see Figure 9). Any two circles S and S' are centrally similar with negative similarity coefficient equal to $-r'/r$ (where r and r' are the radii of the circles), and similarity center O lying on the line AA' joining the centers and such that $OA'/OA = -r'/r$.

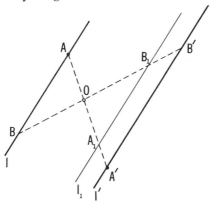

Figure 8

[T] When a transformation $f:\ A \to A'$ is followed by another transformation $g: A' \to A''$, the result is the *composite* transformation $g \circ f:\ A \to A''$; it is usually called the *product* $g \circ f$ of f and g, but in this book it is called the *sum*.

The point O lies inside the segment AA' and is called the *interior center of similarity* of the circles S and S'. To find the interior similarity center of two circles S and S' it is sufficient to draw any two parallel and oppositely directed radii AM and $A'M'$ in these circles; O is the point of intersection of AA' and MM' (Figure 9). If S and S' do not intersect, then their interior similarity center can also be found as the point of intersection of their common interior tangents (Figure 9a); if S and S' are tangent from the exterior, then O is the point of tangency (Figure 9b). Thus any two incongruent circles may be regarded as centrally similar in two ways: the similarity coefficient can be taken equal to r'/r or to $-r'/r$. Two congruent circles are centrally similar in just one way, with similarity coefficient -1. For concentric circles (and only for them) the exterior and interior centers of similarity coincide with each other and with the common center of the circles.

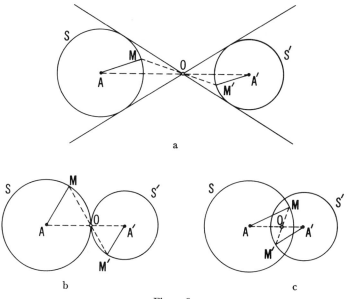

Figure 9

The only fixed point of a central similarity (different from the identity transformation, which can be regarded as the special case of a central similarity with coefficient $k = 1$) is the center O; the fixed lines are all the lines through O.

If the coefficient of a central similarity is -1, then the transformation is a half turn about the similarity center; thus a half turn about a point is a special case of a central similarity. Using this we can generalize Problems 9–11 in Volume One, whose solutions used half turns; to solve the more general problems, one must use more general central similarities.

Thus in Problem 9 one can require that the segment of the desired line lying between the given line and the circle should be divided by the point A in a given ratio m/n; in Problem 10 (a) one can require that the ratio of the lengths of the chords in which the given circles meet the desired line has a given value m/n; in Problem 10 (b) one can require that, when the lengths of the chords cut out of the desired line by the given circles S_1 and S_2 are multiplied by given numbers m and n, then the difference should have a given value; in Problem 11 one can require that the point J divide the segment EF of the chord CD in the ratio m/n. The solutions of these new problems are analogous to the solutions of Problems 9–11; we leave it to the reader to carry out these solutions for himself.

1. Two lines l_1 and l_2 and a point A are given. Pass a line l through A so that the segment BC cut off by l_1 and l_2 satisfies $AB:AC = m:n$.

2. (a) A circle S and a point A on it are given. Find the locus of the midpoints of all the chords through A.

   (b) A circle S and three points A, B, and C on it are given. Draw a chord AX that bisects the chord BC.

3. Let two tangent circles R and S be given. Let l be a line through the point M of tangency, and let this line meet R in a second point A and meet S in a second point B. Show that the tangent to R at A is parallel to the tangent to S at B.

4. Let R and S be two disjoint circles, neither inside the other. Let m be a common tangent to R and S and assume that R and S are both on the same side of m. Let n be another common tangent, with R and S both on the same side of n. Let M be the point of intersection of m and n. Let l be a line through M meeting R in points A and B, and meeting S in points C and D. Finally, let E be the point of tangency of m and R, and let F be the point of tangency of m and S. Prove that:

   (a) Triangle ABE is similar to triangle CDF.

   (b) The ratio of the areas of triangles ABE and CDF is equal to the square of the ratio of the radii of R and S.

   (c) The line determined by the points of intersection of the medians of triangles ABE and CDF passes through the point M.

5. Let $ABCD$ be a trapezoid whose sides AD and BC, extended, meet in a point M; let N be the point of intersection of the diagonals AC and BD. Prove that:

(a) The circles R and S circumscribed about triangles ABM and DCM are tangent.

(b) The circles R_1 and S_1 circumscribed about triangles ABN and CDN are tangent.

(c) The ratio of the radii of R_1 and S_1 is equal to the ratio of the radii of R and S.

6. (a) Using the two parallel sides AB and CD of trapezoid $ABCD$ as bases construct equilateral triangles ABE and CDF. These triangles should each be on the same side of the base (that is, if we regard AB and CD as being horizontal, then either both triangles are constructed above the base, or both are constructed below the base). Prove that the line EF passes through the point of intersection of the extensions of the two nonparallel sides of the trapezoid.

(b) On the parallel sides AB and CD of the trapezoid, construct squares exterior to the trapezoid. Prove that the line joining their center passes through the point of intersection of the diagonals of the trapezoid.

7. Prove that the line joining the midpoints of the two parallel sides of a trapezoid passes through the point of intersection of the extensions of the other two sides, as well as through the point of intersection of the diagonals.

Remark. See also Problem 108 in Section 1 of Chapter I, Volume Three.

8. (a) Two concentric circles S_1 and S_2 are given. Draw a line l meeting these circles consecutively in points A, B, C, D such that $AB = BC = CD$ (Figure 10a).

(b) Three concentric circles S_1, S_2 and S_3 are given. Draw a line l meeting S_1, S_2 and S_3 in order in points A, B, and C such that $AB = BC$ (Figure 10b).

(c) Four concentric circles S_1, S_2, S_3 and S_4 are given. Draw a line l meeting S_1, S_2, S_3 and S_4 respectively in points A, B, C, D such that $AB = CD$ (Figure 10c).

9. (a) Inscribe a square in a given triangle ABC so that two vertices lie on the base AB, and the other two lie on the sides AC and BC.

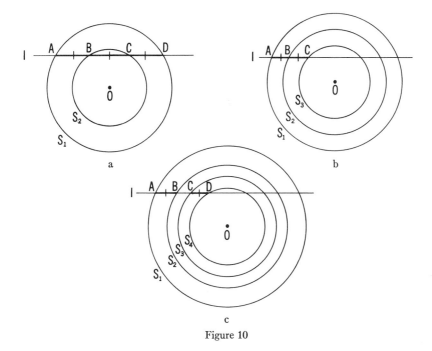

Figure 10

(b) In a given triangle ABC inscribe a triangle whose sides are parallel to three given lines l_1, l_2 and l_3.†

Problem 10(c) is a generalization of Problem 9(b).

10. (a) Two lines l_1 and l_2 are given together with a point A on l_1 and a point B on l_2. Draw segments AM_1 and BM_2 on the lines l_1 and l_2 having a given ratio $AM_1/BM_2 = m$, and through the points M_1 and M_2 pass lines parallel to two given lines l_3 and l_4 (Figure 11). Find the locus of the points of intersection of these lines.

(b) A polygon $A_1A_2\cdots A_n$ varies in such a way that its sides remain parallel to given directions, and the vertices A_1, A_2, $\cdots$, A_{n-1} move on given lines l_1, l_2, $\cdots$, l_{n-1}. Find the locus of the vertex A_n.

(c) In a given polygon inscribe another polygon whose sides are parallel to given lines.†

Problem 189 in Section 5, Chapter I, Volume Three is a substantial generalization of Problem 10 (c).

† By a polygon "inscribed" in a given polygon we mean a polygon, all of whose vertices lie on the sides of the given polygon (with at least one vertex on each side) or on their extensions.

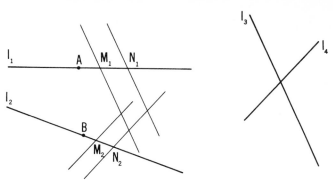

Figure 11

11.   Let a "hinged" parallelogram $ABCD$ be given. More precisely, the lengths of the sides are fixed, and vertices A and B are fixed, but vertices C and D are movable (Figure 12). Prove that, as C and D move, the point Q of intersection of the diagonals moves along a circle.

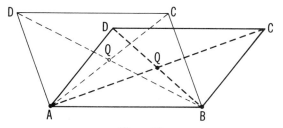

Figure 12

12. (a) Let the inscribed circle S of $\triangle ABC$ meet BC at the point D. Let the escribed circle which touches BC and the extensions of sides AB and AC meet BC at the point E. Prove that AE meets S at the point D_1 diametrically opposite to D.

(b) Construct triangle ABC, given the radius r of the inscribed circle, the altitude $h = AP$ on side BC, and the difference $b - c$ of the two other sides.

13.   Construct a circle S

(a) tangent to two given lines l_1 and l_2 and passing through a given point A;

(b) passing through two given points A and B and tangent to a given line l;

(c) tangent to two given lines l_1 and l_2 and to a given circle S.

See Problem 22 below, and Problems 232 (a), (b), 237 in Section 2, 247 in Section 3, and 271 in Section 5 of Chapter II, Volume Three.

14. (a) Prove that the point M of intersection of the medians of triangle ABC, the center O of the circumscribed circle, and the point H of intersection of the altitudes lie on a line,† and that $HM/MO = 2/1$ (Figure 13a).

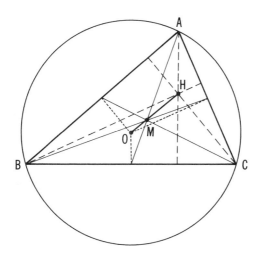

Figure 13a

(b) Prove that the three lines through the midpoints of the sides of a triangle and parallel to the bisectors of the opposite angles meet in a single point.

(c) Prove that the lines joining the vertices of triangle ABC to the points where the opposite sides are tangent to the corresponding escribed circles meet in a single point J. This point is collinear with the point M of intersection of the medians and the center Z of the inscribed circle, and $JM/MZ = 2/1$ (Figure 13b).

15.   Inscribe a triangle ABC in a given circle S, if the vertex A and the point H of intersection of the altitudes are given.

16.   Given a circle S. What is the locus of the points of intersection of the (a) medians, (b) altitudes, of all possible acute angled triangles inscribed in S; of all right angled triangles inscribed in S; of all obtuse angled triangles inscribed in S?

† This line is called the *Euler line* of triangle ABC.

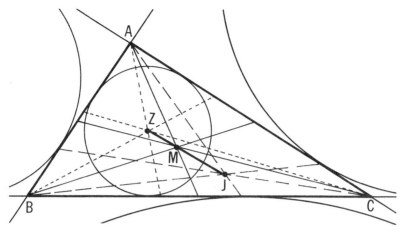

Figure 13b

17. (a) Let H be the point of intersection of the altitudes of triangle ABC, and let M be the point of intersection of the medians. Prove that the circle $\bar{S}$, centrally similar to the circumscribed circle S with similarity center H and similarity coefficient $\frac{1}{2}$, is also centrally similar to S with similarity center M and coefficient $-\frac{1}{2}$. This circle passes through the midpoints A', B', C' of the sides of the triangle, through the feet of the altitudes, and through the midpoints of the segments HA, HB and HC on the altitudes (Figure 14a).†

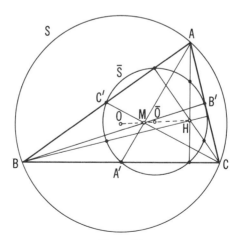

Figure 14a

† This circle $\bar{S}$ is called the *Euler circle* or the *nine point circle* of triangle ABC.

(b) Let J be the point of intersection of the lines joining the vertices of triangle ABC to the points where the opposite sides are tangent to the escribed circles [see Problem 14 (c)] and let M be the point of intersection of the medians of the triangle. Prove that the circle $\bar{S}$ that is centrally similar to the inscribed circle S with similarity center at the point J and similarity coefficient $\frac{1}{2}$ is also centrally similar to S with similarity center M and coefficient $-\frac{1}{2}$. This circle is tangent to the lines $A'B'$, $B'C'$, $C'A'$ joining the midpoints of the sides of triangle ABC, and is also tangent to the three lines joining the midpoints D, E and F of the segments JA, JB and JC (Figure 14b).

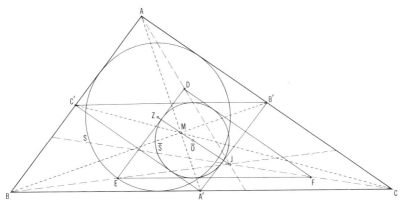

Figure 14b

18. (a) *The centroid of a polygon.* By the *centroid of a segment* we mean its midpoint (Figure 15a). The centroids of the sides of a triangle form a new triangle, centrally similar to the original one with similarity coefficient $-\frac{1}{2}$ and similarity center at the point of intersection of the medians, which point is also called the *centroid of the triangle* (Figure 15b). Prove that the centroids of the four triangles whose vertices coincide with the vertices of an arbitrary given quadrilateral form a new quadrilateral centrally similar to the given one with similarity coefficient $-\frac{1}{3}$; the corresponding center of similarity N (see Figure 15c) is called the *centroid of the quadrilateral.* Analogously, the centroids of the five quadrilaterals whose vertices coincide with the vertices of an arbitrary given pentagon form a new pentagon, centrally similar to the given one with similarity coefficient $-\frac{1}{4}$ and center of similarity at a point that is called the *centroid of the pentagon;* the centroids of the six pentagons whose vertices coincide with the vertices of an arbitrary given hexagon form a new hexagon, centrally similar to the given one with similarity

coefficient $-\frac{1}{5}$ and center of similarity at a point that is called the *centroid of the hexagon*, etc.†

[In other words, the three lines joining the vertices of a triangle to the centroids of the opposite sides meet in a point— the centroid of the triangle—and are divided by it in the ratio $2:1$; the four lines joining each of the vertices of a quadrilateral to the centroid of the triangle formed by the other three vertices meet in a point—the centroid of the quadrilateral—and are divided by it in the ratio $3:1$, and so forth.]

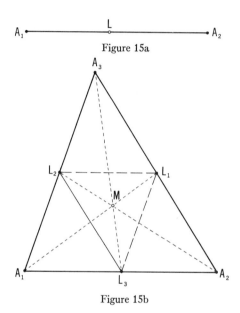

Figure 15a

Figure 15b

(b) *The Euler circle of a polygon inscribed in a circle.* By the *Euler circle of a chord* of a circle S of radius R we mean the circle of radius $R/2$ with center at the midpoint of the chord (Figure 15d). The centers of the three Euler circles of the three sides of a triangle (inscribed in a circle of radius R) lie on a circle of radius $R/2$ (with center at the point of intersection of these three circles), called the *Euler circle of the triangle* (compare Figure 15e with Figure 14a). Prove that the centers of the four Euler circles of the four triangles whose vertices are the vertices of an arbitrary quadrilateral inscribed in a circle S of radius R, lie on a circle of radius $R/2$ (with center at the point of inter-

---

† It is not difficult to prove that the centroid defined here for an n-gon coincides with the physical centroid, or center of gravity, of n equal masses placed at the vertices.

section of these four circles); this circle is called the *Euler circle of the quadrilateral* (Figure 15f). Analogously, the centers of the Euler circles of the five quadrilaterals, whose vertices are the vertices of an arbitrary pentagon, inscribed in a circle S of radius R, lie on a circle of radius $R/2$ (with center at the point of intersection of these five circles) called the *Euler circle of the pentagon*, etc.

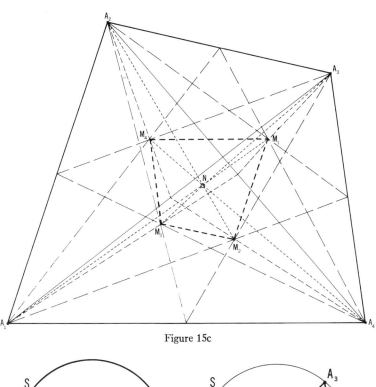

Figure 15c

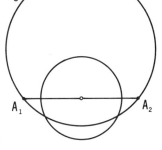

Figure 15d

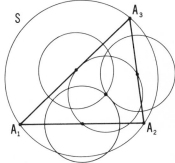

Figure 15e

(c) Prove that the centroid of an n-gon inscribed in a circle [see Problem (a)] lies on the segment joining the center of the circumscribed circle to the center of the Euler circle [see Problem (b)] and divides this segment in the ratio $2:(n-2)$.

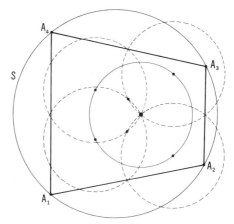

Figure 15f

It is convenient to use central similarities for the solution of *construction problems in a bounded portion of the plane*. Usually in the solution of problems in construction one assumes that the plane is unbounded; thus, for example, one assumes that every line can be extended indefinitely in both directions. Actual constructions, however, are always carried out in a strictly bounded domain—on a sheet of paper or on the classroom blackboard. Therefore in the course of the construction it may happen, for example, that the point of intersection of two lines that enter into the construction lies beyond the limits of the picture, that is, it is inaccessible to us. This circumstance causes us to consider problems in which it is specially stated that the construction must be carried out entirely within some bounded portion of the plane.† The application of

† We note that in *geometry* one doesn't usually insist on the actual completion of a figure in each construction problem: the problem is considered solved if the correct path to its solution has been given (that is, if the construction has been carried out "with the aid of language", in the words of the German geometer J. Steiner (1796–1863)). Therefore the boundedness of the figure is not a real barrier to the solution of construction problems, and so constructions in a bounded portion of the plane must be regarded as belonging to those types of problems whose formulation contains some special condition restricting the possibilities of construction (analogous to constructions in which one is forbidden to use the ruler, or in which one is forbidden to use compasses; for such constructions see Volume Three). In *engineering design* the question of constructions in a bounded portion of the plane has practical significance. Also in *geodesy*—the science of construction and measurement in localities of the earth —such constructions have a serious practical meaning; here the admissible domain for constructions may be bounded by a river, a sea, a mountain, a forest, a swamp, etc.

central similarities enables us to establish the remarkable fact that every construction that can be carried out in the unbounded plane can also be carried out in any portion of it no matter how small! (See Problem 20 below.)

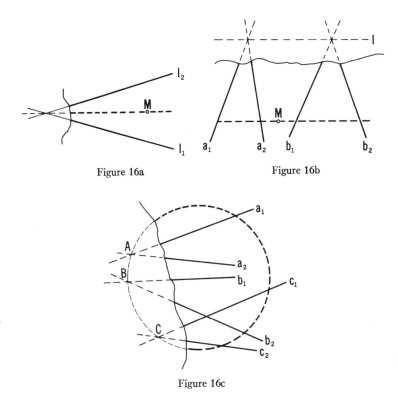

Figure 16a   Figure 16b

Figure 16c

19. (a) Connect a given point M to the "inaccessible" point of intersection of two given lines l_1 and l_2 (or of a given line l and a circle S, or of two given circles S_1 and S_2). (See Figure 16a.)

(b) Through a given point M pass a line parallel to an "inaccessible" line l, two points of which are determined by the intersections of the pairs of lines a_1, a_2 and b_1, b_2 (Figure 16b).

(c) Pass a circle through three "inaccessible", non-collinear points A, B, C determined, respectively, by the pairs of lines a_1 and a_2, b_1 and b_2, c_1 and c_2 (Figure 16c); of course in this problem we are only required to construct that part of the circle lying in the part of the plane accessible to us, or to determine its center and radius. See also Problems 119 (a), (b) and 120 of Section 2, Chapter I of Volume Three.

20.   Show that for any given distribution of points in a bounded portion $\mathcal{K}$ of the plane or even outside of it, and for any magnitudes of given intervals, each problem in construction that can be solved in the whole plane can also be solved without leaving the boundaries of $\mathcal{K}$. [In this connection, if a point A that is given or that is to be constructed lies beyond the boundaries of $\mathcal{K}$, then it is determined by two lines in the domain that intersect at A; an inaccessible line is determined by two of its points, and an inaccessible circle by its center and one point or by its center and radius.]

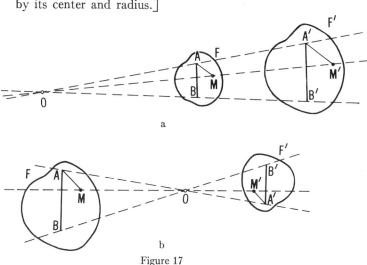

Figure 17

Let F and F' be two centrally similar figures with similarity coefficient k (positive or negative!). (See Figure 17a and b.) In this case corresponding segments in the two figures will be parallel, and their lengths will have the fixed ratio k; this follows from the fact that triangles OAB and $OA'B'$ are similar (since $OA'/OA = OB'/OB = \pm k$). Let us agree to consider the ratio of two parallel segments AB and $A'B'$ positive or negative according to whether they have the same directions (from A to B and from A' to B') or opposite directions; this convention is analogous to the one we introduced earlier for the ratios of segments on the same line. Then in all cases one can say that *corresponding segments in two centrally similar figures are parallel and have a constant ratio, equal to the similarity coefficient.* Let us prove that, conversely, *if to each point of the figure F one can assign some point of the figure F' in such a way that corresponding segments of these two figures are parallel and have a constant* (in magnitude and in sign!) *ratio k, not equal to 1, then F and F' are centrally similar.*

Indeed, choose any point M of F and let M' be the corresponding point of F'; let A and A' be any other pair of corresponding points of

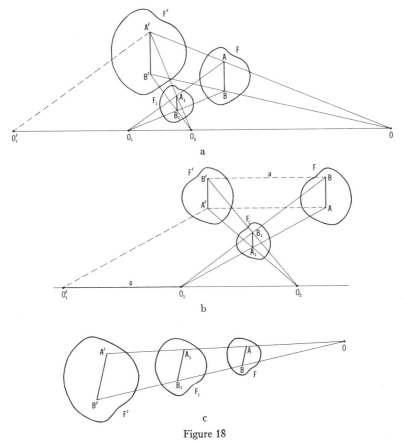

Figure 18

these two figures, and let O be the point of intersection of the lines MM' and AA' (see Figure 17). Since $MA \parallel M'A'$, the triangles OMA and $OM'A'$ are similar; by assumption $M'A'/MA = k$, therefore $OM'/OM = OA'/OA = k$. From this it follows, first, that the point O does not depend on the choice of the pair A, A' (it is the point on the line MM' for which $OM'/OM = k$), and second, that any pair of corresponding points A and A' of the figures F and F' are centrally similar with similarity center O and similarity coefficient k, which is what we were required to prove. If the ratio k of corresponding segments were $+1$, then our reasoning would be invalid, since in this case the lines MM' and AA' would not meet (they would be parallel); in this case the figures F and F' are not centrally similar, but are obtained from one another by a translation (see pages 18 and 19 of Volume One).

Let us now consider the addition of central similarities. Let the figure F_1 be centrally similar to F with similarity center O_1 and similarity coefficient k_1 ; let F' be centrally similar to the figure F_1 with similarity center O_2 and similarity coefficient k_2 (see Figure 18, where for simplicity we have shown the case of positive k_1 and k_2 ; actually in the following reasoning k_1 and k_2 can be taken either positive or negative). In this case corresponding segments in F and F_1 are parallel and have the constant ratio k_1 ; corresponding segments in F_1 and F' will be parallel and will have the constant ratio k_2. From this it follows that corresponding segments in F' and F are parallel and have the constant ratio k_1k_2 (multiplying the left and the right members of the equations $A_1B_1/AB = k_1$ and $A'B'/A_1B_1 = k_2$, we obtain $A'B'/AB = k_1k_2$). But this means that F' is centrally similar to the figure F with similarity coefficient k_1k_2 if $k_1k_2 \neq 1$, and is obtained from F by a translation if $k_1k_2 = 1$. This can also be expressed as follows: *the sum of two central similarities with coefficients k_1 and k_2 is a central similarity with coefficient k_1k_2 if $k_1k_2 \neq 1$, and is a translation if $k_1k_2 = 1$.*†

Given the centers O_1 and O_2 and the coefficients k_1 and k_2 of two central similarities, we now show how to find the center O of the central similarity that is their sum (or, in case $k_1k_2 = 1$, how to find the magnitude and direction of the translation that is their sum).‡ Clearly, if O_1 coincides with O_2 then O coincides with this point also (Figure 18c); therefore we shall assume that the centers O_1 and O_2 are different. The first central similarity leaves the center O_1 in place, while the second carries O_1 into the point O_1' on the line O_2O_1 such that $O_2O_1'/O_2O_1 = k_2$ (Figure 18a, b). Thus the sum of the two transformations carries the O_1 onto O_1'. From this it follows that *if $k_1k_2 = 1$* (Figure 18b), *then the sum of the two transformations is a translation in the direction of the line O_1O_1'* (that is, in the direction of the line O_2O_1, since O_1' lies on the line O_1O_2) *through a distance $a = O_1O_1'$*; since $O_2O_1'/O_2O_1 = k_2$, a can also be represented in the form

$$a = O_2O_1' - O_2O_1 = \frac{O_2O_1' - O_2O_1}{O_2O_1} O_2O_1 = (k_2 - 1)O_2O_1 .$$

If $k_1k_2 \neq 1$ (Figure 18a), *then the desired center O lies on the line O_1O_1', that is, on the line O_1O_2, and $OO_1'/OO_1 = k_1k_2$*. A more convenient expression for the position of the point O can be found. From the relations

---

† Here is another formulation of the same proposition: *two figures F and F' that are each separately centrally similar to a third figure F_1, are either centrally similar to one another or are translates of one another.*

We recommend that the reader try for himself to prove the theorem on the addition of central similarities, starting from the definition of central similarity and without using the fact that two figures whose corresponding segments are parallel and have a constant ratio are centrally similar to one another.

‡ To carry out the reasoning below independently of the signs and magnitudes of the similarity coefficients k_1 and k_2, one must consider directed segments throughout (see text in fine print on pages 20 and 21 of Volume One).

$O_2O_1'/O_2O_1 = k_2$ and $OO_1'/OO_1 = k_1k_2$ it follows that

$$\frac{O_1O_1'}{O_2O_1} = \frac{O_2O_1' - O_2O_1}{O_2O_1} = k_2 - 1 \quad \text{and} \quad \frac{O_1O_1'}{OO_1} = \frac{OO_1' - OO_1}{OO_1} = k_1k_2 - 1;$$

dividing the first of these equations by the second we have

$$\frac{OO_1}{O_2O_1} = \frac{k_2 - 1}{k_1k_2 - 1}, \quad \text{or finally} \quad OO_1 = \frac{k_2 - 1}{k_1k_2 - 1} O_2O_1.$$

Note that along the way we have proved the following important theorem:†

THEOREM on the three centers of similarity. *Let the figure F_1 be centrally similar to the figure F with similarity center O_1, and let it also be centrally similar to the figure F' with similarity center O_2. If O_1 does not coincide with O_2, then the line O_1O_2 passes through the center of similarity O of the figures F and F'* (Figure 18a), *or is parallel to the direction of the translation carrying F into F'* (Figure 18b). *If O_1 coincides with O_2, then O_1 is the center of similarity also for F and F'* (Figure 18c).

If O_1 is different from O_2 then the line O_1O_2 is called the *axis of similarity of the three figures* F, F_1 and F'; if O_1 coincides with O_2 then this point is called the *center of similarity of the figures* F, F_1 any F'.

Usually the theorem on the three centers of similarity is formulated somewhat less precisely as follows: *the three centers of similarity of three pairwise centrally similar figures lie on a line.*‡

As an example consider the case of three circles S_1, S_2 and S_3. In the general case, when no two of them are congruent, each pair of circles has two centers of similarity, the exterior center and the interior center, so that in all there are six centers of similarity lying on four axes of similarity (Figure 19). If two of the circles are congruent, then they do not have an exterior center of similarity. Thus there are five centers of similarity lying on the four axes of similarity; if all three circles are congruent to one another, then there are three centers of similarity and three axes of similarity. Also, all the axes of similarity are distinct if the centers of the three circles do not lie on a line; if the centers are collinear, then

---

† An entirely different proof of this theorem is outlined in the last paragraph of Chapter II, Volume Three.

‡ The case when the three centers of similarity coincide is included in this formulation, which is valid in all cases when three centers of similarity exist. The lack of precision mentioned in the text consists in the fact that here we have excluded from consideration the case when two of the three figures F, F_1 and F' are congruent (are obtained from one another by a translation). In this connection see Section 2 of Chapter I, Volume Three.

the axes of similarity all coincide, in which case it may happen that three of the centers of similarity coincide, so that one of the four axes of similarity of the three circles becomes a center of similarity.†

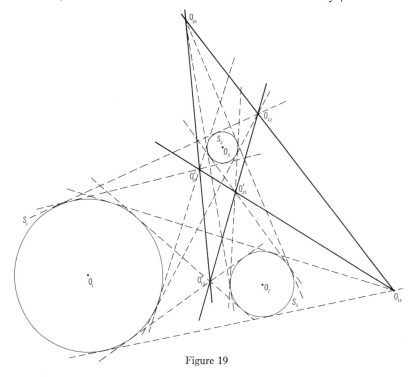

Figure 19

There is an elegant stereometric proof of the theorem on the three centers of similarity. Let us designate the plane in which the three figures F, F_1 and F' lie by the letter π. Extend the figures F, F_1 and F' to be spatial figures $\bar{F}$, $\bar{F}_1$ and $\bar{F}'$, pairwise centrally similar with the same similarity centers O_1, O_2 and O (Figure 20); ‡ if F' is not centrally similar to F but is obtained from F by a translation, then $\bar{F}'$ is obtained from $\bar{F}$ by the same translation. Let A be an arbitrary point of $\bar{F}$ that does not lie in the plane π, and let A_1 and A' be the corresponding points in $\bar{F}_1$ and $\bar{F}'$. Then the line A_1A passes through O_1, A_1A' passes through O_2, and AA' passes through O (or is parallel to the direction of the translation carrying F into F'). Hence, if O_1 and O_2 coincide, then the lines A_1A and $A'A_1$ also coincide; therefore the line AA' will also coincide with AA_1 and A_1A', which means that the point O, their point of intersection with the plane π, coincides with O_1 and O_2. If $O_1 \neq O_2$, then the plane through A, A_1 and A' meets π in a line l that passes through all three centers of similarity O_1, O_2 and O, or passes through O_1 and O_2 and is parallel to the direction of the translation carrying F into F'.

† The reader is urged to draw figures for himself, illustrating all the possible cases.

‡ The definition and properties of central similarities (dilatations) in space are analogous to the definition and properties of central similarities in the plane.

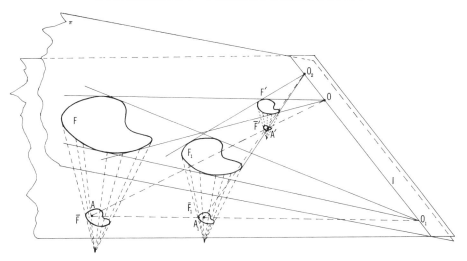

Figure 20

21. Let the circle S be tangent to each of the circles S_1 and S_2. Prove that the line joining the points of tangency passes through a center of similarity of S_1 and S_2 (the exterior center, if S is tangent to S_1 and S_2 in the same sense, that is, internally tangent to both or externally tangent to both; the interior center otherwise).

This problem occurs in another connection in Section 1, Chapter II, Volume Three (see Problem 212).

22. Use the theorem on the three centers of similarity to derive a new solution to Problem 13c.

23. (a) Let the circles S_1 and S_2 be externally tangent at the point M_1 (that is, they touch at this point and neither lies inside the other); let the circles S_2 and S_3 be externally tangent at M_2, and let S_3 and S_1 be externally tangent at M_3 (Figure 21a). Let A_1 be an arbitrary point of S_1, and let A_2 be the second point of intersection of the line A_1M_1 with S_2; let A_3 be the second point of intersection of A_2M_2 with S_3, and let A_4 be the second point of intersection of A_3M_3 with S_1. Prove that A_1 and A_4 are diametrically opposite points of S_1.†

Generalize the result of this exercise to the case of an arbitrary *odd* number of tangent circles.

† If, for example, $A_2 = M_2$, then the line A_2M_2 must be replaced by the tangent to S_2 at M_2, so that $A_3 = M_2$ also.

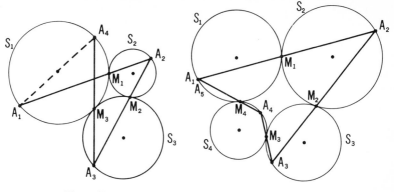

Figure 21a   Figure 21b

(b) Let the circles S_1 and S_2 be externally tangent at the point M_1; let the circles S_2 and S_3 be externally tangent at M_2; let the circles S_3 and S_4 be externally tangent at M_3; finally, let S_4 and S_1 be externally tangent at M_4 (Figure 21b). Let A_1 be an arbitrary point of S_1, and let A_2 be the second point of intersection of the line A_1M_1 with S_2; let A_3 be the second point of intersection of A_2M_2 with S_3; let A_4 be the second point of intersection of A_3M_3 with S_4, and let A_5 be the second point of intersection of A_4M_4 with S_1. Prove that A_1 and A_5 coincide.†

Generalize the result of this problem to the case of an arbitrary *even* number of tangent circles.

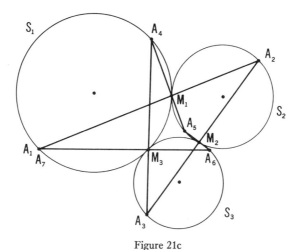

Figure 21c

† See the preceding footnote.

(c) With the notations of part (a), let A_5 be the second point of intersection of line A_4M_1 with S_2, let A_6 be the second point of intersection of A_5M_2 with S_3, and let A_7 be the second point of intersection of A_6M_3 with S_1 (Figure 21c). Show that the point A_7 coincides with A_1.†

Generalize this result to the case of an arbitrary number of tangent circles.

(d) In what ways are the results of the preceding three parts, (a), (b), (c), changed if we do not insist that in each the circles be *externally* tangent to one another?

24. Let the circles R and S be externally tangent at the point M, and let t be a common tangent to these two circles, meeting them in the points N and P respectively (Figure 22). Let A be an arbitrary point of R, and let B be the second point of intersection of the line AM with S. Through N draw a line parallel to PB, and let this line meet R in the point C. Prove that A and C are diametrically opposite points on the circle R.

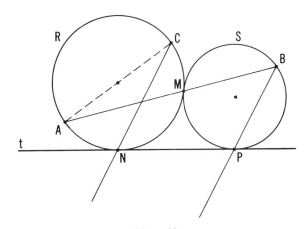

Figure 22

25. Let ABC be a given triangle, and let k be a line parallel to BC which meets sides AC and AB in points K and L respectively; let m be a line parallel to CA meeting sides BA and BC in points M and N; let p be a line parallel to AB meeting sides CB and CA in points P and Q. Prove that the points of intersection of AB and KN, of BC and MQ, and of CA and PL, if they all exist, are collinear.

† See footnote on p. 31.

26. (a) Let P be an arbitrary point in the plane, and let K, L, M be the points that are symmetric to P with respect to the mid-points D, E, F of the sides AB, BC, CA of a given triangle ABC. Prove that the segments CK, AL and BM meet in a common point Q which is the midpoint of each of them.

(b) Let the point P of part (a) move around a circle S. What path does the point Q describe?

27.   Let M, N and P be three points on the sides AB, BC and CA of a triangle ABC (or on their extensions). Prove that

(a) the three points M, N and P are collinear if and only if

$$\frac{AM}{BM} \cdot \frac{BN}{CN} \cdot \frac{CP}{AP} = 1$$

(*theorem of Menelaus*);†

(b) the three lines CM, AN and BP are concurrent or parallel if and only if

$$\frac{AM}{BM} \cdot \frac{BN}{CN} \cdot \frac{CP}{AP} = -1$$

(*theorem of Ceva*).†

The theorems of Menelaus and Ceva are presented in a different connection in Section 2, Chapter I of Volume Three [see Problem 134 (a), (b)]; many applications of these important theorems are indicated there.

28. (a) Using the assertions of Problems 18 (a) and 14 (a) derive a new solution to Problem 34 (a) of Section 1, Chapter II, Volume One (p. 47).

† Note that one must prove two theorems: 1) if the points M, N and P lie on one line, then

$$\frac{AM}{BM} \cdot \frac{BN}{CN} \cdot \frac{CP}{AP} = 1$$

(the necessity of the condition), and 2) if

$$\frac{AM}{BM} \cdot \frac{BN}{CN} \cdot \frac{CP}{AP} = 1,$$

then the points M, N and P lie on one line (the sufficiency of the condition); Problem 27 (b) is to be interpreted similarly. For the concept of the sign of the ratio of segments, see page 12.

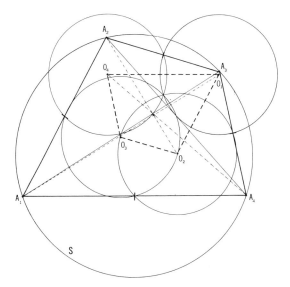

Figure 23

(b) Let A_1, A_2, A_3, A_4 be four points on a circle S; let O_1, O_2, O_3, O_4 be the centers of the Euler circles [see Problem 17 (a)] of the triangles $A_2A_3A_4$, $A_1A_3A_4$, $A_1A_2A_4$ and $A_1A_2A_3$. Show that the quadrilateral $O_1O_2O_3O_4$ is centrally similar to the quadrilateral $A_1A_2A_3A_4$ with similarity coefficient $\frac{1}{2}$ (Figure 23).

[In other words, if the points A_1, A_2, A_3 and A_4 all lie on a circle, then the four segments joining each of these points to the center of the Euler circle of the triangle formed by the other three meet in one point and are divided by this point in the ratio 2:1.]

29. Let A_1, A_2, A_3 and A_4 be points on the circle S; let H_4, H_3, H_2 and H_1 be the points of intersections of the altitudes of triangles $A_1A_2A_3$, $A_1A_2A_4$, $A_1A_3A_4$ and $A_2A_3A_4$. From the eight points A_1, A_2, A_3, A_4, H_1, H_2, H_3 and H_4, select all triples with the property that their subscripts are distinct, and consider all the triangles having these triples of points for vertices (for example, $\triangle A_1A_2A_4$ and $\triangle A_1H_2A_4$ are admissible, while $\triangle A_1A_3H_3$ is not, since A_3 and H_3 have the same subscript). There are $(8 \cdot 6 \cdot 4)/6 = 32$ of them. For each of these triangles an Euler circle can be constructed [see Problem 17 (a)]. Prove that

(a) among these 32 circles only eight are distinct;

(b) these eight circles are all congruent and meet in a single point;

(c) they can be divided into two sets such that the centers of the four circles in either set are centrally similar to the four points A_1, A_2, A_3, A_4 with similarity coefficient $\frac{1}{2}$ and to the four points H_1, H_2, H_3, H_4 with similarity coefficient $-\frac{1}{2}$.

2. Spiral similarity and dilative reflection. Directly similar and oppositely similar figures.

Let F_1 be the figure centrally similar to the figure F with center O and *positive* similarity coefficient k. Rotate F_1 through an angle α about the point O into the position F' (Figure 24). The transformation carrying F into F' is called a *spiral similarity*.† Thus, a spiral similarity is characterized by two magnitudes: the *similarity coefficient k* and the *angle of rotation α*. The point O is called the *center* of the spiral similarity.

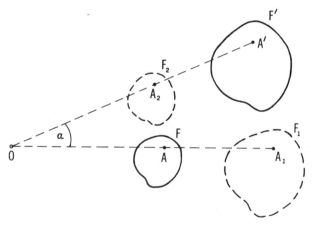

Figure 24

Spiral similarities can also be realized as follows: first rotate F about the center O through the angle α into the position F_2, and then perform the central similarity with similarity center O and similarity coefficient k, carrying F_2 into F'. In other words, *a spiral similarity with center O, rotation angle α and similarity coefficient k is the sum of a central similarity with center O and similarity coefficient k and a rotation about O through the angle α, taken in either order.*‡ It follows from this that if F' is obtained

† Such transformations are sometimes called *rotary dilatations* (this name is customary, for example, in mathematical crystallography).

‡ It follows from this that a spiral similarity is a *similarity transformation* in the sense of the definition on page 6 of the introduction to this volume, because a central similarity is a similarity transformation, and a rotation is an isometry.

from F by a spiral similarity with center O, rotation angle α and similarity coefficient k, then, conversely, F can be obtained from F' by a spiral similarity (with the same center O, rotation angle $-\alpha$ and similarity coefficient $1/k$); thus we can speak of figures obtained from one another by spiral similarities.

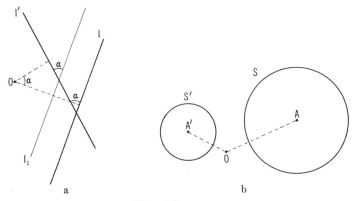

a   b

Figure 25

A spiral similarity carries each line l into a new line l' (Figure 25a). To construct l' one must first construct the line l_1, centrally similar to l with similarity center O and similarity coefficient k, and then rotate l_1 about O through the angle α into the position l'. The lines l and l' form an angle α (see Volume One, page 30). A circle S is carried by a spiral similarity into a new circle S' (Figure 25b); the center A' of S' is the image of the center A of S under the spiral similarity, the radius r' of S' is equal to kr, where r is the radius of S and k is the similarity coefficient.

A rotation about a point is a special case of a spiral similarity (with similarity coefficient $k = 1$). This allows us to generalize some of the problems of Volume One, whose solutions used rotations; in the solutions of the more general problems one must use spiral similarities instead of rotations. Thus, with the conditions of Problem 18 of Section 2, Chapter I, Volume One, the equilateral triangle can be replaced by a triangle similar to an arbitrarily given triangle; with the conditions of Problem 20 one can require that the chords that are cut off on the desired lines l_1 and l_2 by the circles S_1 and S_2 have an arbitrarily given ratio. The solutions of these generalizations of Problems 18 and 20 are analogous to the solutions of the original problems; we leave it to the reader to carry them out for himself.

Another special case of a spiral similarity is a central similarity (a central similarity with coefficient k is a spiral similarity with rotation angle $\alpha = 0°$ if $k > 0$, and with angle $\alpha = 180°$ if $k < 0$). Corresponding to this one can generalize some of the problems of Section 1 of this chapter: in the solutions of the new problems one must use spiral similarities instead of central similarities. Thus, for example, with the conditions of Problem 1 one can ask for points B and C such that the segments AB and AC do not lie on one line l, but on two lines l and m, passing through the point A and forming a given angle α (see also Problem 34, which is a generalization of Problem 13 (c) of Section 1).

The only *fixed point* of a spiral similarity (other than the identity transformation, which can be regarded as a spiral similarity with rotation angle 0° and similarity coefficient 1) is the similarity center O. A spiral similarity that is not a central similarity (that is, a spiral similarity with rotation angle different from 0° and 180°), has no *fixed lines*.

30. (a) In a given triangle ABC inscribe a triangle PXY (the point P is given on side AB) similar to a given triangle LMN.

   (b) In a given parallelogram $ABCD$ inscribe a parallelogram similar to a given parallelogram $KLMN$.

31.   Let S_1 and S_2 be two given intersecting circles, and let A be one of the points of intersection; let a line through A be given, and let it meet S_1 and S_2 in M_1 and M_2 respectively; let N be the point of intersection of the tangents to the circles at M_1 and M_2; through the centers O_1 and O_2 of the circles draw lines parallel to M_1N and to M_2N, and let J be their point of intersection (Figure 26). Prove that the line JN always passes through a certain point, and that the segment JN always has the same length (independent of the choice of the line M_1AM_2).

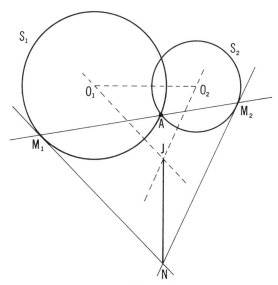

Figure 26

32. (a) Construct a quadrilateral $ABCD$ that can be inscribed in a circle and whose sides have given lengths $AB = a$, $BC = b$, $CD = c$, $DA = d$.

   (b) Construct a quadrilateral $ABCD$, given the sum of the opposite angles B and D, and given the lengths of the sides $AB = a$, $BC = b$, $CD = c$, $DA = d$.

   Problem (a) is a special case of Problem (b).

33. Let R and S be two intersecting circles, and let M be one of the points of intersection; let l be any line through M, and let l meet R and S in points A and B.† As l varies, find the locus of:

   (a) the point Q that divides the segment AB in a given ratio $AQ/QB = m/n$;

   (b) the vertex C of the equilateral triangle ABC constructed on the segment AB;

   (c) the endpoint P of the segment OP, laid off from a fixed point O and equal to, parallel to, and having the same direction as the segment AB.‡

34. Construct a circle S tangent to two given lines l_1 and l_2 and cutting a given circle $\bar{S}$ at a given angle α. [By *the angle between two circles* we mean the angle between their tangent lines at the point of intersection. The angle between tangent circles is zero; circles that do not intersect do not form any angle.]

Let the figure F' be obtained from the figure F by a spiral similarity with rotation angle α and similarity coefficient k (Figure 27). Let AB and $A'B'$ be any two corresponding segments in these two figures. In this case $A'B'/AB = k$ (because in Figure 27, where F_1 is obtained from F by a rotation through the angle α and F' is obtained from F_1 by a central similarity with coefficient k, we have $A_1B_1 = AB$; $A'B'/A_1B_1 = k$), and the angle between the segments $A'B'$ and AB is equal to α (because the angle between AB and A_1B_1 is equal to α and $A'B' \parallel A_1B_1$). Therefore, *corresponding segments in the figures F' and F have a constant ratio k and form a constant angle α.* Let us now prove that, conversely, *if to each point of F there corresponds a point of F' in such a manner that corresponding segments in these figures have a constant ratio k and form a*

---

† If, for example, l is tangent to S, then the point $B = M$ (compare footnote † on page 31).

‡ In other words, in Problem 33 (c) we are required to find the locus of the endpoints of the vectors $\overrightarrow{OP} = \overrightarrow{AB}$, laid off from a fixed point O.

constant angle α (the segments of the figure F become parallel to the corresponding segments of the figure F' when they are rotated through an angle α in a definite direction), *then F and F' are related by a central similarity.* Indeed, let M and M' be any two corresponding points of F and F' (Figure 27). On the segment MM' construct a triangle $MM'O$ such that $OM'/OM = k$ and $\sphericalangle MOM' = \alpha$. † (If $\alpha > 180°$, then we construct the triangle with $\sphericalangle MOM' = 360° - \alpha$.)

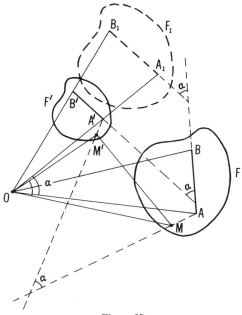

Figure 27

The spiral similarity with center O, rotation angle α and coefficient k carries M into M'; let us prove that it carries each point A of F into its corresponding point A' of F'. Consider the triangles OMA and $OM'A'$.

† One must first make a preliminary construction of a triangle T with vertex angle α and with the ratio of the adjacent sides equal to k. Since T is similar to MOM', T determines the angles at the base MM' of triangle MOM'.

Triangle $MM'O$ is constructed on that side of the segment MM' such that the directed rotation through the angle α carrying the line OM into the line OM' coincides with the directed rotation through the angle α that makes the segments of F parallel to the corresponding segments of F'.

The point O can also be found as the intersection of the circular arc each point of which subtends an angle α with the segment MM', and a second circle—the locus of points the ratio of whose distances to M and M' is equal to k (see, for example, *Introduction to Geometry,* by H. S. M. Coxeter, Chapter 6 on the circle of Apollonius; also see Problem 257 in Section 4 of Chapter II, Volume Three).

In these triangles $OM'/OM = M'A'/MA$ (since $OM'/OM = k$ by construction and $M'A'/MA = k$ by hypothesis) and $\angle OMA = \angle OM'A'$ (because the angle between OM and OM' is equal to α by construction and the angle between MA and $M'A'$ is equal to α by hypothesis; see Volume One, page 33); hence they are similar. From this it follows that $OA'/OA = k$; also, $\angle AOA' = \angle MOM' = \alpha$ (because $\angle AOM = \angle A'OM'$). But this means that our spiral similarity carries A into A'.

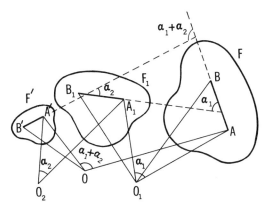

Figure 28

Now it is easy to answer the question: what is the sum of two spiral similarities? Let the figure F_1 be obtained from the figure F by a spiral similarity with center O_1, similarity coefficient k_1 and rotation angle α_1,† and let the figure F' be obtained from the figure F_1 by a spiral similarity with center O_2, similarity coefficient k_2 and rotation angle α_2 ; let AB, A_1B_1 and $A'B'$ be corresponding segments in these three figures (Figure 28). Then, $A_1B_1/AB = k_1$, and the segments AB and A_1B_1 form an angle α_1 ; $A'B'/A_1B_1 = k_2$, and the segments A_1B_1 and $A'B'$ form an angle α_2 . Therefore,

$$\frac{A'B'}{AB} = \frac{A_1B_1}{AB} \cdot \frac{A'B'}{A_1B_1} = k_1k_2,$$

and the segments AB and $A'B'$ form an angle $\alpha_1 + \alpha_2$ (see the footnote on page 32 of Volume One). Thus corresponding segments in the figures F and F' have a constant ratio k_1k_2 and form a constant angle $\alpha_1 + \alpha_2$. By what has been proved above this means that the figure F' is obtained from F by a spiral similarity with rotation angle $\alpha_1 + \alpha_2$ and with similarity coefficient k_1k_2, or, in case $\alpha_1 + \alpha_2 = 360°$ and $k_1k_2 = 1$,

† Here and in what follows angles of rotation are always measured in one fixed direction, for example, the counter-clockwise direction.

by a translation. Thus *the sum of two spiral similarities with similarity coefficients k_1 and k_2 and rotation angles α_1 and α_2 is a new spiral similarity with similarity coefficient $k_1 k_2$ and rotation angle $\alpha_1 + \alpha_2$; in the excluded case, when $k_1 k_2 = 1$ and $\alpha_1 + \alpha_2 = 360°$,†* the sum of the two spiral similarities is a translation.*‡

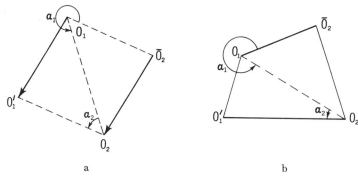

a   b

Figure 29

Let us find the center O of the spiral similarity (or the direction and magnitude of the translation) represented by the sum of two given spiral similarities. If their centers O_1 and O_2 coincide, then, clearly, O coincides with this point. Now assume that O_1 does not coincide with O_2. The sum of the two spiral similarities carries the point O_1 into a point O_1', and, in fact, the second rotation alone carries O_1 into this point (since the first rotation leaves O_1 fixed); it is not difficult to construct the point O_1'. The sum of the two transformations carries some point $\bar{O}_2$ into the point O_2, and, in fact, the first rotation alone carries this point into O_2 (since the second rotation leaves O_2 fixed); $\bar{O}_2$ is easy to construct. Hence, if $k_1 k_2 = 1$ and $\alpha_1 + \alpha_2 = 360°$, then the segments $O_1 O_1'$ and $\bar{O}_2 O_2$ are equal, parallel and have the same direction (Figure 29a); each of these segments gives the magnitude and direction of the translation that is the sum of our two spiral similarities. If the sum of the two spiral similarities is another spiral similarity, then this new spiral similarity carries the segment $O_1 \bar{O}_2$ into $O_1' O_2$ (Figure 29b).

† More precisely, $\alpha_1 + \alpha_2$ is a multiple of 360° (see the footnote of page 34 of Volume One).

‡ We urge the reader to attempt to find an independent proof of the theorem on the addition of spiral similarities, using only the definition of such transformations, without using the fact that if the corresponding segments in two figures have a constant ratio and form a constant angle then they may be obtained from one another by spiral similarity.

We now show *how to construct the center O of the spiral similarity carrying a given segment AB into another given segment $A'B'$* (in our case $O_1\bar{O}_2$ into $O_1'O_2'$).† If the angle between the segments is 180° or 360° and the segments are not equal, then the spiral similarity becomes a central similarity; in this case O is the point of intersection of the lines AA' and BB' (Figure 30a). Suppose now that the angle between the segments is not a multiple of 180°, that is, the segments AB and $A'B'$ are not parallel; let P denote the point of intersection of AB and $A'B'$ (Figure 30b). Then the circles circumscribed about the triangles $AA'P$ and $BB'P$ pass through the rotation center O: indeed, $\angle AOA' = \angle APA'$ (the angle of rotation AOA' is equal to the angle APA' between the segments AB and $A'B'$); analogously $\angle BOB' = \angle BPB'$;‡ therefore *the center O can be found as the second point of intersection of the circles circumscribed about triangles $AA'P$ and $BB'P$* (see Figure 30b).

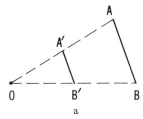

a

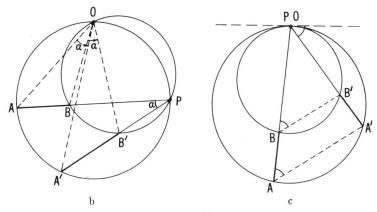

b   c

Figure 30

† To find the center O one can also carry out an auxiliary construction of a triangle with vertex angle $\alpha_1 + \alpha_2$ [or $360° - (\alpha_1 + \alpha_2)$] whose adjacent sides have the ratio k_1k_2, and in this manner determine the base angles of triangle $O_1'O_1O$ (see the footnote on page 40).

‡ See the second footnote on page 50 of Volume One.

If these two circles are tangent at P (Figure 30c), that is, if $\angle PAA' = \angle PBB'$ (these angles are both equal to the angle between the line $PA'B'$ and the common tangent to the circles at P), then $O = P$ (since from the similarity of triangles PAA' and PBB' we have $PA'/PA = PB'/PB$).

Let us note too, that the center of the spiral similarity carrying AB into $A'B'$ coincides with the center of the spiral similarity carrying AA' into BB'; indeed if $\angle AOA' = \angle BOB' = \alpha$, and $OA'/OA = OB'/OB = k$, then $\angle AOB = \angle A'OB' = \beta$ (in Figure 31, $\beta = \alpha + \angle A'OB$) and $OA/OB = OA'/OB' = l$. It follows that *the center O can also be found as the point of intersection of the circles circumscribed about triangles ABQ and $A'B'Q$, where Q is the point of intersection of the lines AA' and BB'* (or as the point of intersection of the lines AB and $A'B'$, if $AA' \parallel BB'$; this last case, pictured in Figure 30c, will hold when the circles circumscribed about triangles $AA'P$ and $BB'P$ are tangent at P).

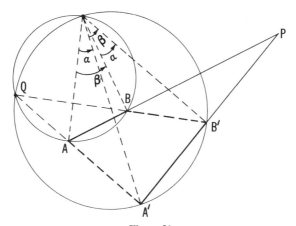

Figure 31

35.   Four lines l_1, l_2, l_3 and l_4 are given in the plane, no three of which meet in a point and no two of which are parallel. Prove that the circles circumscribed about the four triangles formed by these lines meet in a common point (Figure 32).

See also Problem 62 (a) in Section 1, Chapter II (page 77). A far reaching generalization of Problem 35 will be found in Problem 218 (a) in Section 1, Chapter II of Volume Three.

36.   Let S_1 and S_2 be two circles with centers O_1 and O_2, meeting in a point A. Consider a fixed angle α with vertex at A. Let M_1 and M_2 be the points of intersection of the sides of this angle with the circles S_1 and S_2, and let J be the point of intersection of the lines O_1M_1 and O_2M_2 (Figure 33).

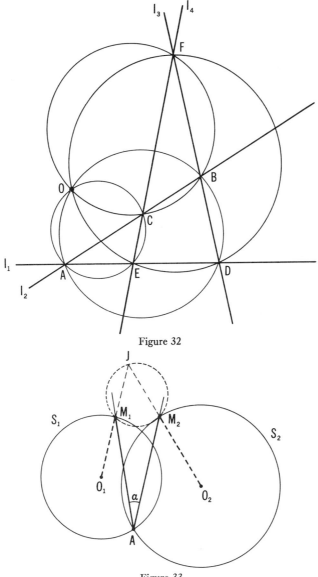

Figure 32

Figure 33

(a) Show that as the angle is rotated about the point A, the circle circumscribed about triangle M_1M_2J always passes through a certain fixed point O.

(b) Find the locus of all the points O of part (a), corresponding to all possible angles α.

37.   Construct an n-gon, knowing the vertices M_1, M_2, $\cdots$, M_n of the n triangles constructed on its sides as bases and similar to n given triangles. Under what conditions will the problem have no solution, and when will it have more than one solution?

This problem is a generalization of Problem 21, Section 2, Chapter I, of Volume One, page 37.

38.   Let S_1 and S_2 be two circles intersecting in points M and N. Let A be an arbitrary point on S_1, let B be the second point of intersection of the line AM with the circle S_2, let C be the second point of intersection of the line BN with S_1, let D be the second point of intersection of CM with S_2, and let E be the second point of intersection of DN with S_1.†

(a) Prove that the distance AE does not depend on the choice of the initial point A on S_1, but only on the two circles S_1 and S_2.

(b) How must S_1 and S_2 be situated for E to coincide with A?

39.   Let S_1, S_2 and S_3 be three circles, each of which intersects both of the other two, and let A_1 be an arbitrary point on S_1.

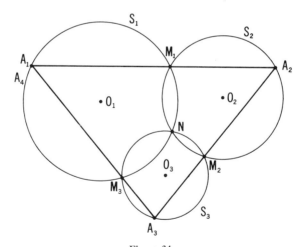

Figure 34a

† Several exceptional cases can occur. First, for example, if $A = M$, then the line AM is replaced by the tangent to S_1 at M.

Second, if A is so placed then AM is tangent to S_2, then B is taken to coincide with M, and therefore C coincides with M also; then just as in the first case the line CM must be replaced by the tangent to S_1 at M. Compare the footnote to Problem 23 (a), page 31.

(a) Assume that the three circles have a common point of intersection N, and let M_1, M_2, M_3 be the second points of intersection of S_1 and S_2, S_2 and S_3, S_3 and S_1, respectively (Figure 34a). Let A_2 be the second point of intersection of the line A_1M_1 with S_2, let A_3 be the second point of intersection of A_2M_2 with S_3, and let A_4 be the second point of intersection of A_3M_3 with S_1.† Prove that A_4 coincides with A_1.

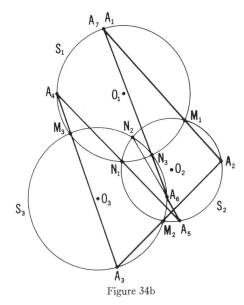

Figure 34b

(b) Now assume that there is no common point of intersection of the three circles, and let S_1 and S_2 meet in the points M_1 and N_1, let S_2 and S_3 meet in the points M_2 and N_2, let S_3 and S_1 meet in the points M_3 and N_3 (Figure 34b). Let A_2 be the second point of intersection of A_1M_1 with S_2, let A_3 be the second point of intersection of A_2M_2 with S_3, let A_4 be the second point of intersection of A_3M_3 with S_1, let A_5 be the second point of intersection of A_4N_1 with S_2, let A_6 be the second point of intersection of A_5N_2 with S_3, and let A_7 be the second point of intersection of A_6N_3 with S_1.‡ Prove that A_7 coincides with A_1.

Generalize the results of Problems 39 (a) and (b) to the case of an arbitrary number of pairwise intersecting circles.

† If, for example, A_1M_1 is tangent to S_2 at M_1 then A_2 is considered to coincide with M_1; on the other hand, if the point A_2 coincides with M_2 then, by the line A_2M_2, we mean the tangent to S_2 at M_2 (compare the footnote to Problem 38).

‡ See the footnotes to Problems 38 and 39 (a).

40. Let S_1 and S_2 be two circles intersecting in two points M and N, let l be a line, and let A be an arbitrary point on S_1.

(a) Let l meet S_1 in points K and L, and let it meet S_2 in points P and Q (Figure 35a). Let B be the second point of intersection of the line AM with the circle S_2. Let C be the second point of intersection of the line $KC \parallel BP$ with S_1, let D be the second point of intersection of the line CN with S_2, and let E be the second point of intersection of the line $LE \parallel DQ$ with S_1.† Prove that E coincides with A.

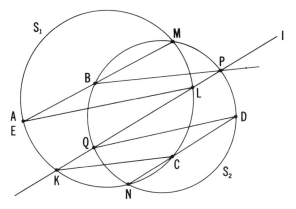

Figure 35a

(b) Let the line l be tangent to S_1 and S_2 at the points K and P (Figure 35b). Let B be the second point of intersection of the line AM with S_2, let C be the second point of intersection of the line $KC \parallel BP$ with S_1, let D be the second point of intersection

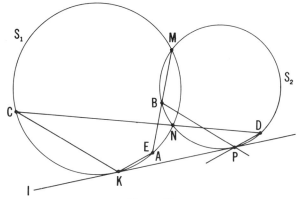

Figure 35b

† See the footnotes to Problems 38 and 39 (a).

of CN with S_2, and let E be the second point of intersection of the line $KE \parallel DP$ with S_1.† Prove that E and A coincide.

Let the figure F_1 be centrally similar to the figure F with similarity center O and (positive!) similarity coefficient k; let F' be the figure obtained from F_1 by reflection in some line l through O (Figure 36). In this case we say that F' is obtained from F by a *dilative reflection* with *similarity coefficient* k; the point O and the line l are called the *center* and the *axis* of the dilative reflection. The same dilative reflection can be realized in a different manner: first perform a reflection in the line l, carrying F into the figure F_2, and then perform the central similarity with center O and coefficient k, carrying F_2 into F'. In other words, *a dilative reflection is the sum of a central similarity with center O and a reflection in a line l through O*, performed in either order.‡

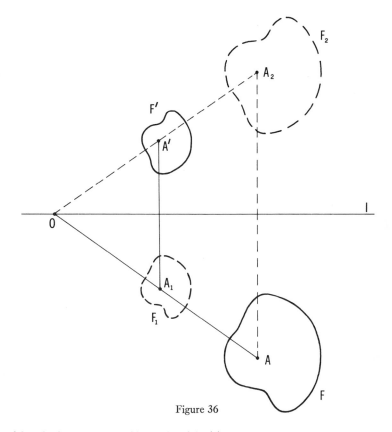

Figure 36

† See the footnotes to Problems 38 and 39 (a).

‡ It follows that a dilative reflection is a *similarity transformation* in the sense of the definition given in the introduction to this volume (since it is the sum of a similarity transformation and an isometry).

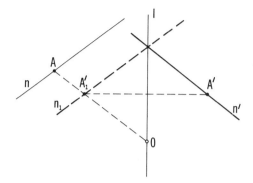

Figure 37a

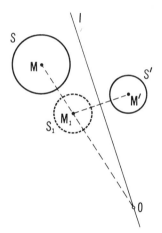

Figure 37b

Clearly, if the figure F' is obtained from F by a dilative reflection, then conversely, F may be obtained from F' by a dilative reflection (with the same center O and axis l, but with similarity coefficient $1/k$); this permits us to speak of figures obtained from one another by dilative reflections. A dilative reflection takes a line n into a new line n' (Figure 37a), and takes a circle S into a new circle S' (Figure 37b).

The only *fixed point* of a dilative reflection (that is not simply a reflection in the line l, which can be regarded as a dilative reflection with similarity coefficient $k = 1$) is the center O of the transformation. The only *fixed lines* of a dilative reflection (that is not a reflection in a line) are the axis l of the transformation and the perpendicular to l at O.

41. Let a line l, a point A on it, and two circles S_1 and S_2 be given. Construct a triangle ABC in which the line l is the bisector of the angle A, the vertices B and C lie respectively on the circles S_1 and S_2 and the ratio of the sides AB and AC has a given value $m:n$.

42. Construct a quadrilateral $ABCD$ whose diagonal bisects the angle A, given

(a) sides AB and AD, the diagonal AC, and the difference of the angles B and D;

(b) sides BC and CD, the ratio of sides AB and AD, and the difference of angles B and D;

(c) sides AB and AD, the diagonal AC, and the ratio of sides BC and CD.

43. Solve Problem 42 replacing the condition that the diagonal AC be the bisector of angle A by the more general condition that angles BAC and DAC have a given difference γ.

44. Two lines l_1 and l_2 are given in the plane. With each point M in the plane we associate a new point M' in the following manner:

(a) Let m be a line through M such that the segment PQ included on this line between l_1 and l_2 is bisected by M.† Let P' be the orthogonal projection of P on l_2, let Q' be the orthogonal projection of Q on l_1, and let M' be the midpoint of $P'Q'$ (Figure 38a).

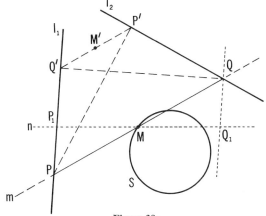

Figure 38a

† To construct m it is sufficient to pass an arbitrary line n through M, cutting l_1 in P_1, and to lay off $MQ_1 = MP_1$; then we will have $QQ_1 \parallel l_1$ (Figure 38a).

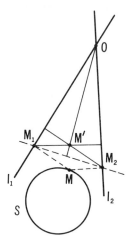

Figure 38b

(b) Let O be the point of intersection of l_1 and l_2, let M_1 and M_2 be the orthogonal projections of M onto l_1 and l_2, and let M' be the point of intersection of the altitudes (that is, the orthocenter) of triangle OM_1M_2 (Figure 38b).

Suppose now that M describes a circle S; what will be the locus described by M'?

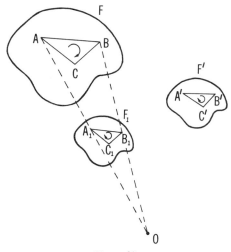

Figure 39a

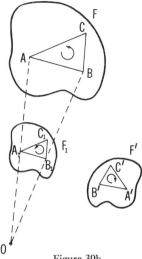

Figure 39b

Just as we distinguished between directly congruent and oppositely congruent figures (see Volume One, page 61), we shall distinguish two kinds of similar figures in the plane. Two similar figures F and F' will be called *directly similar* if the figure F_1, centrally similar to F with similarity coefficient equal to the ratio of corresponding segments in F' and F, is directly congruent to F' (Figure 39a); on the other hand, if F_1 is oppositely congruent to F' then we shall say that F and F' are *oppositely similar* (Figure 39b). We shall say that two figures are similar, without further qualification, only when it does not matter to us if they are directly or oppositely similar. To decide whether two given similar figures F and F' are directly similar or oppositely similar it is sufficient to consider any three non-collinear points A, B, C of F, and the corresponding points A', B', C' of F'. If F and F' are directly similar, then the (similar) triangles ABC and $A'B'C'$ have the same direction around the boundary; otherwise they have opposite directions around the boundary (see Figure 39a and b).

Let us now prove the following two important theorems.

THEOREM 1. *Any two directly similar figures in the plane can be made to coincide by means of a spiral similarity or by a translation.*†

† It follows from this theorem that any two directly similar but non-congruent figures in the plane can be made to coincide by means of a spiral similarity (since if the figures could be made to coincide by means of a translation, then they would be congruent).

The proof of this theorem hardly differs at all from the proof of Theorem 1 of Section 2, Chapter II, Volume One (pp. 61–63): First of all it is easy to show that any two segments AB and $A'B'$ can be made to coincide by means of a spiral similarity with rotation angle equal to the angle between the segments, and with similarity coefficient equal to the ratio of the segments; the only exception occurs when the segments are equal, parallel, and have the same direction, in which case they can be carried into each other by a translation (see page 40, and Volume One, pp. 18–19, where a more general statement is proved concerning arbitrary figures). Now by means of a spiral similarity or a translation, carry some segment AB of the figure F into the corresponding segment $A'B'$ in the figure F', directly similar to F. It is not difficult to show that the entire figure F is taken into F'; the proof of this is entirely analogous to the concluding part of the proof of Theorem 1, Section 2, Chapter II, Volume One.

If F and F' can be made to coincide by means of a spiral similarity with center O, then O is called the *rotation center* (or sometimes the *similarity center*) of F and F'. To find the rotation center O of two directly similar figures F and F', it is necessary to choose a pair of corresponding segments AB and $A'B'$ in these figures. If the lines AB and $A'B'$ intersect in a point P, and if AA' and BB' intersect in Q, then O is the other point of intersection of the circles circumscribed about triangles PAA' and PBB', or is the second point of intersection of the circles circumscribed about triangles QAB and $QA'B'$ (see pages 43–44). If $AB \parallel A'B'$, then O coincides with Q; if $AA' \parallel BB'$, then O coincides with P. Finally, if $AB \parallel A'B'$ and $AA' \parallel BB'$, then the segments AB and $A'B'$ are equal, parallel and have the same direction; in this case the figures F and F' are carried onto each other by a translation and there is no center of rotation.

THEOREM 2. *Any two oppositely similar figures in the plane can be made to coincide by means of a dilative reflection or by means of a glide reflection.*†

The proof of this theorem is very similar to the proof of Theorem 2 of Section 2, Chapter II, Volume One (pp. 64–65). Let A and B be any two points of F, and let A' and B' be the corresponding points of the figure F', oppositely similar to F. We shall prove that there is a dilative reflection (or a glide reflection) carrying the segment AB into the segment $A'B'$. Indeed, let l be the axis of a dilative (or glide) reflection carrying AB into $A'B'$ (Figure 40). Translate $A'B'$ into the new position $A\bar{B}$. The segment A_1B_1, obtained by reflecting AB in l, is centrally

---

† It follows from this theorem, in particular, that any two oppositely similar, but not congruent, figures in the plane can be obtained from each other by means of a dilative reflection (since if the figures can be made to coincide by a glide reflection, then they are congruent).

similar to $A'B'$ or it is obtained from $A'B'$ by a translation; consequently the segments A_1B_1 and $A\bar{B}$ are parallel to $A'B'$. Since A_1B_1 is the reflection of AB in l and is parallel to $A\bar{B}$, it follows that l is parallel to the bisector l_0 of angle $\bar{B}AB$.

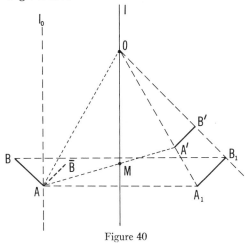

Figure 40

Further, suppose that $A'B' \neq AB$, and that O is the center of the dilative reflection carrying AB into $A'B'$. Then l is an angle bisector of triangle AOA_1, and hence of triangle AOA'. Therefore the line AA' crosses l in a point M such that

$$\frac{A'M}{MA} = \frac{OA'}{OA} = \frac{OA'}{OA_1} = \frac{A'B'}{A_1B_1}$$

or

$$\frac{A'M}{MA} = \frac{A'B'}{AB}.$$

Thus if we know the segments AB and $A'B'$ then we can construct the line l: it is parallel to the bisector of angle $\bar{B}AB$ and divides the segment AA' in the ratio $A'B'/AB$. [This construction works even when $A'B' = AB$; see the proof of Theorem 2 of Section 2, Chapter II, Volume One.] A reflection in l carries the segment AB into the segment A_1B_1, parallel to $A'B'$. A_1B_1 can be carried into $A'B'$ by means of a central similarity with center O at the point of intersection of $A'A_1$ and l (because $OA'/OA_1 = OA'/OA = A'M/MA = A'B'/AB$) or by a translation in the direction of the line l. Thus we have found the dilative (or glide) reflection whose existence we wished to show. The concluding part of the proof of Theorem 2 is almost word for word the same as the concluding part of the proof of Theorem 2 of Chapter II, Volume One, and therefore we suppress it here.

The results of Theorems 1 and 2 can be formulated in the following general proposition:

Any two similar figures in the plane can be carried into each other by means of a spiral similarity, or a dilative reflection, or a translation, or a glide reflection.†

This proposition can be made the basis for a *definition* of similarity transformation in the plane (see Volume One, pp. 68–70). Namely, a similarity transformation is any transformation of one of the following four types:
1) spiral similarities (including central similarities and rotations about points);
2) dilative reflections;
3) translations;
4) glide reflections (including reflections in lines).

It is natural to call transformations of types 1) and 3) *direct similarities* (they carry directly similar figures into one another), and to call transformations of types 2) and 4) *opposite similarities* (they carry oppositely similar figures into one another).

Theorems 1 and 2 enable us to generalize Problems 45–47 of Section 2, Chapter II, Volume One, by replacing the requirement that the segments AX and BY in Problem 45, AX, BY and CZ in Problem 46, BP, PQ and QC in Problem 47 be congruent, by the requirement that they have a given ratio (ratios). The solutions of these more general problems are analogous to the solutions of the original problems; we recommend that the reader solve them for himself.

From Theorems 1 and 2 it follows that the properties common to all similarity transformations are precisely those properties that are common to spiral similarities, dilative reflections, translations, and glide reflections (see the definition of similarity transformation, above). Thus, for example, *every similarity transformation carries lines into lines and circles into circles.*‡ Also, *every similarity transformation has either a fixed*

† In particular, any two similar but not congruent figures can be carried into each other by means of a spiral similarity or by means of a dilative reflection.

‡ Incidentally, this property of similarity transformations also follows from their definition as those transformations of the plane that preserve the ratio of distances between points. Indeed, the line AB can be defined as the set of those points M for which the largest of the three distances AM, BM, and AB is equal to the sum of the other two, that is, either

$$\frac{AM}{AB} + \frac{BM}{AB} = 1, \quad \text{or} \quad \frac{BM}{AM} + \frac{AB}{AM} = 1, \quad \text{or} \quad \frac{AM}{BM} + \frac{AB}{BM} = 1$$

(Figure 41a). But a similarity transformation that carries the points A and B into points A' and B' carries the set of points M described above into the set of those points M' for which either

$$\frac{A'M'}{A'B'} + \frac{B'M'}{A'B'} = 1, \quad \text{or} \quad \frac{B'M'}{A'M'} + \frac{A'B'}{A'M'} = 1, \quad \text{or} \quad \frac{A'M'}{B'M'} + \frac{A'B'}{B'M'} = 1,$$

that is, the line AB is carried into the line $A'B'$.

point or a fixed line, since among the four types of similarity transformation listed above the only ones that do not have a fixed point are the translations and the glide reflections, but they have fixed lines. Finally, *every oppositely similar transformation has at least one fixed line*. Indeed, a dilative reflection has exactly two fixed lines, while a glide reflection has exactly one fixed line, unless the glide reflection reduces to a reflection in a line, in which case it has infinitely many fixed lines.

It is interesting to note that the first of the properties of similarities listed above is CHARACTERISTIC for similarity transformations, that is, it can be used as the basis for their (descriptive!) definition:† *any transformation of the plane that carries each line into a line and each circle into a circle is necessarily a similarity transformation.* In other words, any such transformation necessarily preserves the ratio of distances between points: if the points A, B, C, D are carried by the transformation into points A', B', C', D' then

$$\frac{C'D'}{A'B'} = \frac{CD}{AB}.$$

This fact underlines the basic role of similarity transformations in elementary geometry, the study of properties of lines, circles, and of figures bounded by line segments and circular arcs.

To prove the above italicized proposition‡ we must show that a transformation which carries straight lines into straight lines and circles into circles carries each pair of points A, B, d units apart, into a pair of points A', B', d' units apart, where d' *depends only on* d and not on the particular choice of the points A, B! This can be shown, for example, by the following reasoning.

We first note that any transformation of the plane onto itself that carries lines into lines must necessarily carry *parallel lines into parallel lines*. Indeed, if our transformation were to carry two parallel lines k, l into lines k', l' intersecting at some point M (Figure 42), then the point that is carried into the point M would have to lie on both k and l, and there is no such point.

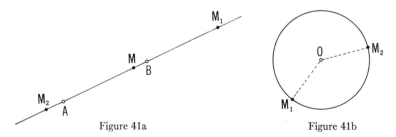

Figure 41a   Figure 41b

In an analogous manner a circle with center O can be defined as the set S of points with the property that if M_1 and M_2 are any two points of S then $OM_1 = OM_2$, that is, $OM_1/OM_2 = 1$ (Figure 41b). But a similarity transformation carrying the point O into some point O' also carries the set S described above into a set S' of points with the property that if M'_1 and M'_2 are any two points of S' then $OM'_1/OM'_2 = 1$. In other words, the transformation carries circles with center O onto circles with center O'.

† Compare pp. 69–70 in Volume One.

‡ See also Section 4 of Chapter II, Volume Three.

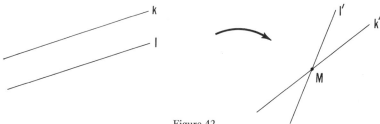

Figure 42

Further, it is clear that if a transformation carries lines into lines and circles into circles, then two circles that meet in two points, or a line and a circle that meet in two points, are carried into two circles that meet in two points, or into a line and a circle that meet in two points. Also, two tangent circles (that is, two circles having exactly one common point), or a circle and a tangent line, are carried into two tangent circles, or into a circle and a tangent line. Finally, two circles, or a circle and a line, that have no points in common are carried into two circles, or a circle and a line, having no points in common. It follows from all this that *two congruent circles*, that is, two circles having parallel common tangents, *are carried into congruent circles* (Figure 43). Further, if our transformation carries a circle S into a circle S' then it must carry the center O of S into the center O' of S'. Indeed, the point O is characterized by the fact that any two circles through O and tangent to S are congruent. Likewise, O' is characterized by the fact that any two circles through it that are tangent to S' are congruent (Figure 44). But we have already seen that congruent circles are carried into congruent circles, and tangent circles are carried into tangent circles, and therefore O must be carried into O'.

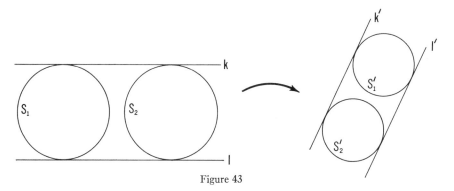

Figure 43

But now it is clear that if the distances AB and CD are equal, that is, if the circle S_1 with center A and radius AB and the circle S_2 with center C and radius CD are congruent, then the transformation must carry the points A, B, C, D into points A', B', C', D' such that the circle S'_1 with center A' and radius $A'B'$ and the circle S'_2 with center C' and radius $C'D'$ are congruent (Figure 45). Hence *if the segments AB, CD have the same length, so do the transformed segments $A'B'$, $C'D'$*, which was to be proved.

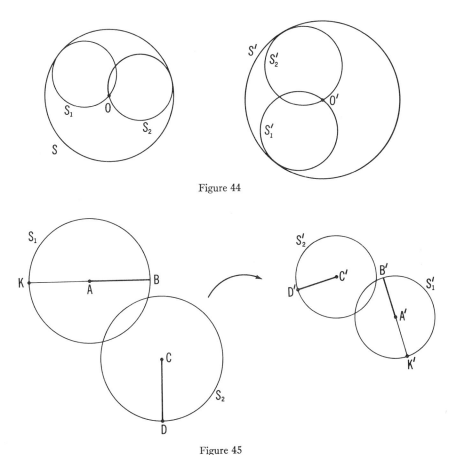

Figure 44

Figure 45

The final part of the proof is fairly standard. In geometry this sort of reasoning occurs rather frequently. First note that if three points K, A, B on a line are such that $KA = AB$, then our transformation will carry them into three points K', A', B' on a line such that $K'A' = A'B'$ —this follows from the fact that if two segments are equal then our transformation carries them into two equal segments (see Figure 45). It follows at once from this that if the ratio $CD/AB = m/n$ is *rational* (m and n are positive integers), then our transformation carries the points A, B, C, D into points A', B', C', D' such that $C'D'/A'B' = CD/AB$ ($=m/n$). Indeed, the condition $CD/AB = m/n$ is equivalent to the existence on the line AB of $n - 1$ points A_1, A_2, $\cdots$, A_{n-1}, and on the line CD of $m - 1$ points C_1, C_2, $\cdots$, C_{m-1}, such that

$$AA_1 = A_1A_2 = \cdots = A_{n-1}B = CC_1 = C_1C_2 = \cdots = C_{m-1}D;$$

but our transformation clearly carries these points into points A'_1, A'_2, $\cdots$, A'_{n-1}; B'_1, B'_2, $\cdots$, B'_{m-1}, of which the first $n - 1$ lie on the line $A'B'$, and the final

$m - 1$ points lie on the line $C'D'$, such that

$$A'A'_1 = A'_1A'_2 = \cdots = A'_{n-1}B' = C'C'_1 = C'_1C'_2 = \cdots = C'_{m-1}D'$$

(Figure 46a).

Further, it only remains to note that if $MN > PQ$, then N lies outside the circle S with center M and radius PQ, that is, from N we can draw two tangents to S; it follows from this that our transformation carries the points M, N, P, Q, with $MN > PQ$, into points M', N', P', Q' with $M'N' > P'Q'$ (Figure 47). Therefore, if the ratio of the distances AB and CD is not rational, and if

$$\frac{m}{n} < \frac{CD}{AB} < \frac{m+1}{n},$$

where m and n are positive integers, then there are points E and F on the line CD such that

$$\frac{CE}{AB} = \frac{m}{n}, \quad \frac{CF}{AB} = \frac{m+1}{n} \quad \text{and} \quad CE < CD < CF.$$

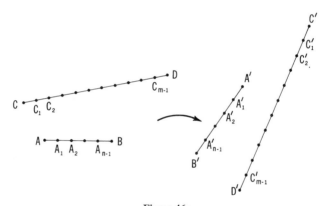

Figure 46a

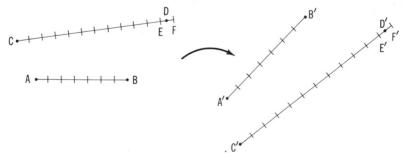

Figure 46b

Thus our transformation carries the points A, B, C, D, E, F into points A', B', C', D', E', F' such that E' and F' lie on the line $C'D'$ and

$$\frac{C'E'}{A'B'} = \frac{m}{n}, \quad \frac{C'F'}{A'B'} = \frac{m+1}{n} \quad \text{and} \quad C'E' < C'D' < C'F'$$

(Figure 46b). It follows from this that

$$\frac{m}{n} < \frac{C'D'}{A'B'} < \frac{m+1}{n}.$$

In this inequality and in the corresponding inequality for CD/AB the denominator n can be arbitrarily large; we conclude that $C'D'/A'B' = CD/AB$.

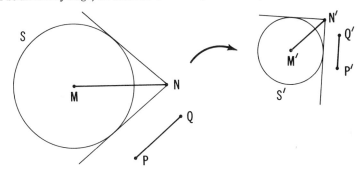

Figure 47

45. (a) Let $ABCD$ and $A_1B_1C_1D_1$ be two arbitrary squares. Assume that the perimeters of the two squares are traversed in the same direction, that is, as we go from A to B to C to D, and from A' to B' to C' to D', then the perimeters of the two squares are either both traversed in the clockwise direction or are both traversed in the counterclockwise direction. Prove that then the midpoints of the segments AA_1, BB_1, CC_1, and DD_1 also form a square, or else these four points all coincide.

Does the conclusion remain valid if the perimeters of the two squares are traversed in opposite directions?

(b) Let ABC and $A_1B_1C_1$ be two equilateral triangles. Using each of the segments AA_1, BB_1 and CC_1 as a base we construct new equilateral triangles AA_1A^*, BB_1B^* and CC_1C^*. Assume that the perimeters of the five triangles ABC, $A_1B_1C_1$, AA_1A^*, BB_1B^* and CC_1C^* are all traversed in the same direction (either all clockwise or all counterclockwise). Prove that the three points A^*, B^* and C^* form the vertices of an equilateral triangle, or else coincide.

Does the conclusion remain valid if we do not require that all five triangles are traversed in the same direction?

46. (a) Let $ABCD$ and $MNPQ$ be two squares. Prove that if the perimeters of these two squares are traversed in the same direction, then

$$AM^2 + CP^2 = BN^2 + DQ^2.$$

Does this conclusion remain valid if we replace the two squares by two similar rectangles? If we do not require that the perimeters be traversed in the same direction?

(b) Let $ABCDEF$ and $MNPQRS$ be two regular hexagons. Prove that if the perimeters are traversed in the same direction, then

$$AM^2 + CP^2 + ER^2 = BN^2 + DQ^2 + FS^2.$$

47. (a) Construct a rectangle whose sides pass through four given points A, B, C and D, and whose diagonal has a given length.

(b) Construct a quadrilateral $ABCD$ with given angles and diagonals.

(c) Let four lines passing through a single point be given. Construct a parallelogram whose sides have given lengths, and whose vertices lie on the four given lines.

48. (a) A point M and two lines l_1 and l_2 are given in the plane. Construct a triangle ABC such that M divides the side AB in a given ratio $BM/MA = k$, and such that l_1 and l_2 are the perpendicular bisectors of BC and CA.

(b) Two points M and N and a line l are given in the plane. Construct a triangle ABC such that M and N divide the sides AB and BC in given ratios $BM/MA = k_1$ and $CN/NB = k_2$, and such that l is the perpendicular bisector of AC.

CHAPTER TWO

Further Applications of Isometries and Similarities

1. Systems of mutually similar figures

In this section we shall consider some systems of mutually similar figures having interesting properties. First we consider the simpler case of systems of congruent figures.

Let F and F' be two congruent figures in the plane. In Section 2 of Chapter II, Volume One, it was shown that these figures can be made to coincide by means of a rotation or translation, and on the basis of this it was asserted that rigid motions within the plane are limited to rotations and translations.† There, however, we did not consider the intermediate positions occupied by the figure during the course of the motion; instead our entire attention was directed to the initial and final positions. Now, however, we shall be interested in the intermediate positions of a moving figure; these positions form a system of mutually congruent figures.‡ There are infinitely many different systems of this sort, cor-

† Here and in what follows by the words "congruent figures" we shall always mean "directly congruent figures" (see p. 61 of Volume One); two oppositely congruent figures cannot in general be made to coincide by means of a rigid motion taking place entirely within the plane.

‡ We emphasize that we are interested only in the set of different positions of moving figures, not in the process of motion itself; thus we shall not at all consider the velocity or acceleration of the individual points. Though considerations from mechanics often simplify proofs of geometric theorems, we do not have the space to treat this distinctive chapter of geometry (so-called *kinematic geometry*) here. [We note that the basic Theorems 1 and 2 of this section could also have been proved by the methods of mechanics.]

responding to the different possible methods of moving a figure from the position F to the position F'; we shall consider only some simple examples of such systems.

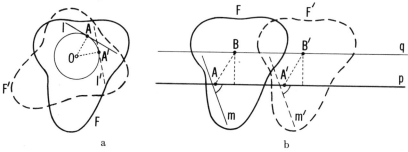

a   b

Figure 48

Let the figure F be *rotated* about the point O; that is, let it move so that a certain point O, which we shall assume belongs to the figure, remains fixed (Figure 48a). In this case each point A of F describes a circle with center O (since the distance OA remains constant); each line l of F is either always tangent to a fixed circle with center O, or always passes through O (since the distance from O to l remains constant). The point O is the rotation center for any two positions of the figure.

Let us now consider the *translation* of a figure in the direction of a given line p (Figure 48b), that is, a motion of the figure in which some line p remains fixed (slides along itself). Then each point B of the figure describes a line parallel to p (because the distance from B to p remains constant); each line m, not parallel to p, is moved in such a way that it remains parallel to its original position (because the angle between m and p does not change); each line q parallel to p slides along itself (because the distance from it to p does not change). Any two positions of F can be obtained from each other by a translation in the direction of p.

If a figure F is moved in the plane in such a way that throughout the motion two *parallel* lines p and q of F always pass through two given points A and B of the plane, then the line p slides along itself (since the angle between p and the segment AB does not change: the sine of this angle is equal to the ratio of the distance between p and q to the length of the segment AB).[T] Thus we are dealing with a translation of the figure, already considered above (see Figure 48b). The situation is more complicated if two *non-parallel* lines of the figure always pass through two given points; such a motion can arise in the following manner: stick two pins into the plane and attach an angle to the figure, then move the angle so that its two sides always touch the pins. Here we have the following theorem:

[T] Here we think of the figure F and the lines p and q as sliding over the plane, while the points in the underlying plane, such as A and B, do not move.

THEOREM 1. *If a figure F is moved in the plane in such a manner that two given non-parallel lines p and q of F always pass through two given points A and B in the plane, then every other line of F either always passes through some given point in the plane, or is always tangent to some circle in the plane.*

PROOF: Let F and F_1 be two positions of F, let p and q, p_1 and q_1 be the corresponding positions of the lines p and q, and let O and O_1 be the points of intersection of p and q, p_1 and q_1 (Figure 49a). The points O and O_1 lie on the arc of the circle S constructed on the chord AB and subtending an angle equal to the angle between p and q.† Let l and l_1 be

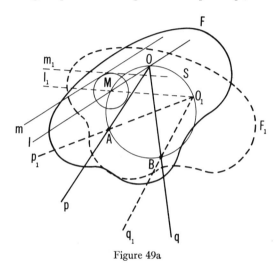

Figure 49a

two positions of some line l that passes through O, and let M and M_1 be the points of intersection of l and l_1 with the circumference S. Since $\angle MOA = \angle M_1O_1A$ (because the angle between the lines l and p does not change), it follows that arc $AM = $ arc AM_1, that is, $M_1 = M_2$. Thus we have shown that all positions of the line l pass through the same point M.

Now let m be a line that does not pass through O. Pass a line l through the point O parallel to m. As we have seen, all positions of the line l pass through the same point M. Since the distance between the lines m and l remains constant, all positions of the line m must touch the circle with center at M and radius equal to the distance from l to m. This completes the proof of the theorem.

Now suppose that the *figure F moves in the plane in such a manner that two non-parallel lines m and n of the figure are always tangent to*

† See footnote ‡ on page 50 of Volume One.

two given circles S_1 *and* S_2 (Figure 49b). Through the centers A and B of the circles S_1 and S_2 pass lines p and q parallel to m and n respectively. The distance between m and p is equal to the radius of S_1 ; since this distance does not change during the motion of the figure, the line p at all times passes through the fixed point A. Similarly, the line q passes at all times through B. Hence we can apply Theorem 1; we see that also in such a movement *every line of the figure F is either always tangent to some fixed circle, or always passes through some given point.*

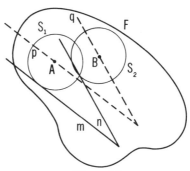

Figure 49b

If the figure F moves in the plane in such a manner that some two points A and B in the figure describe *parallel* lines p and q, then all positions of the segment AB are parallel to one another (because the sine of the angle between AB and p does not change: it is equal to the ratio of the distance between p and q to the length of the segment AB). Hence any two positions of the figure can be obtained from each other by a translation in the direction of p; thus we have a translation of the figure, which we have considered earlier (see Figure 48b). The situation is more complicated if some segment AB of the figure moves with its endpoints on two *non-parallel* lines. Here we have the following theorem:

THEOREM 2. *If a figure F moves in the plane in such a manner that two of its points A and B trace out lines p and q that intersect in a point O, then there exists a circle S attached to the figure F, all points of which trace out lines passing through O.*

PROOF: Let F and F_1 be two positions of the figure F and let AB and A_1B_1 be the corresponding positions of the segment AB (Figure 50). Construct the circle S passing through the points A, B and O. We shall consider this circle to be attached to the figure F, and S_1 will denote the position of this circle when F has reached the position F_1 : clearly S_1 also passes through O (because arc AB of circle S is equal to arc A_1B_1 of circle S_1 which is $2 \sphericalangle AOB$).† Let M be an arbitrary point of S and let

† See footnote ‡ on page 50 of Volume One.

M_1 be the corresponding point of S_1. From the congruence of F and F_1 it follows that the arcs AM and A_1M_1 are equal; this means that the inscribed angles AOM and A_1OM_1 on these arcs are also equal. But this means that line OM_1 coincides with line OM. Consequently, each point M on the circumference S moves along a line passing through O, as was to be proved.

Note also that the center N of S moves along a circle with center O and radius equal to the radius R of the circle S; this follows from the fact that all positions of S pass through the point O and, therefore, the distance ON remains equal at all times to R.

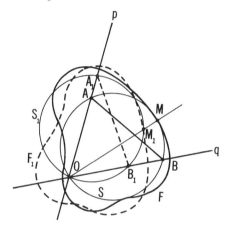

Figure 50

49. Construct a triangle congruent to a given triangle and whose sides (a) pass through three given points; (b) are tangent to three given circles.

> This problem occurred in a different connection in Volume One, Chapter I, Section 1 [see Problem 7 (b) on page 18].

50. (a) The hypotenuse of a right triangle slides with its endpoints on two perpendicular lines. Find the locus described by the vertex at the right angle.

   (b) The longest side of an isosceles triangle with vertex angle of 120° slides with its endpoints on the sides of an angle of 60°. Find the locus described by the vertex at the largest angle.

51. In the plane are given two perpendicular lines l_1 and l_2 and a circle S. Construct a right triangle ABC with a given acute angle α, whose vertices A and B lie on l_1 and l_2, and whose vertex C at the right angle lies on S.

Now let F and F_1 be two similar figures in the plane.† In Chapter I, Section 2, we saw that F can be carried into F' by a spiral similarity; at that time, however, we did not consider the intermediate positions occupied by the figure during its motion from the position F to the position F'. Now we shall consider the system of mutually similar figures made up of all the positions of the figure as it moves from some position F to some other position F' in such a way that at all times it is similar to the first position.

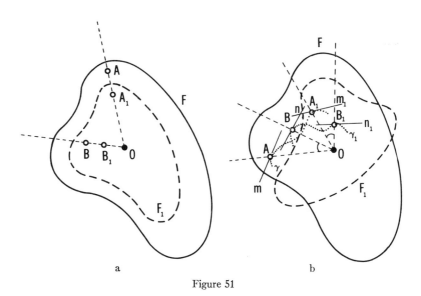

a   b

Figure 51

First suppose that F *moves in such a way that it is at all times similar to its original position and, in addition, so that some point O of the figure does not move at all.* If at the same time some other point A of the figure describes a circle with center O then the distance between O and A does not change; consequently F remains at all times congruent (and not merely similar) to its original position, and all points of F describe circles with center O (see Figure 48a, p. 64). If some point A of F describes a line passing through O, then every other point B also describes a line passing through O (because angle BOA cannot change, as F remains similar to its original position); such a motion of the figure means that at all times it can be taken back to its original position by a central similarity with similarity center O (Figure 51a). Now let us suppose that *the point A of F describes an arbitrary curve* γ; we shall show that *then*

† Here and in the future by the word "similar" we shall mean "directly similar" (see page 53).

every other point B (except for the point O) *describes a curve similar to* γ (Figure 51b). Indeed, let F be the initial position and let F_1 be any other position of the figure F; let A, B and A_1, B_1 be the corresponding positions of the points A and B. Since all positions of the figure are similar to the initial position, and since the point O does not move, it follows that the triangles OAB and OA_1B_1 are similar; consequently, $\angle B_1OA_1 = \angle BOA$ and $OB_1/OA_1 = OB/OA$. Denote the angle BOA by α and the ratio OB/OA by k; the last equations show that a spiral similarity with center O, similarity coefficient k and rotation angle α carries B_1 into A_1. Since F_1 was an arbitrary position of the figure this means that this spiral similarity carries the entire curve γ', described by the point B, into the curve γ, described by the point A. But if two curves can be carried onto one another by a spiral similarity then they are similar, as was to be proved.

In the same way it can be shown that *if some line* m *of the figure* F, *not passing through* O, *is tangent at all times to some curve* γ, *then any other line* n *of the figure* (not passing through O) *is at all times tangent to a curve similar to* γ (see Figure 51b). This curve is obtained from γ by the spiral similarity with center O that carries m into n. In particular, if some line m of the figure F, not passing through O, always passes through some given point M, then any other line n of the figure (not passing through O) will pass at all times through some constant point (this point will not be the same for all lines).

Let us note also that if the figure F moves so that one of its points O remains fixed, then O is the rotation center for any two positions of the figure. Indeed, from the similarity of triangles AOB and A_1OB_1 (Figure 51b) it follows that triangles AOA_1 and BOB_1 are also similar ($\angle AOA_1 = \angle BOB_1$, since $\angle AOA_1 = \angle AOB + \angle BOA_1$ and $\angle BOB_1 = \angle A_1OB_1 + \angle BOA_1$; moreover $OA_1/OA = OB_1/OB$, since $OB_1/OA_1 = OB/OA$). But this means that the figure F_1 is obtained from F by a spiral similarity with center at O, rotation angle A_1OA and similarity coefficient OA_1/OA.

Conversely, if F is moved in such a manner that it always remains similar to its original position and such that any two positions of F have the same rotation center O, then the point O, considered as a point of F, does not move (because the rotation center is a fixed point under a central similarity transformation).

52. Let A be one of the points of intersection of two circles S_1 and S_2. Through A we pass an arbitrary line l and a fixed line l_0, intersecting S_1 and S_2 for the second time in points M_1, M_2 and N_1, N_2; let M_1M_2P be an equilateral triangle constructed on the segment M_1M_2 and let Q be the point of intersection of the lines M_1N_1 and M_2N_2 (Figure 52). Prove that when the line l is rotated around A,

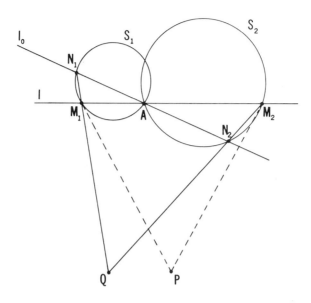

Figure 52

(a) the vertex P of the triangle M_1M_2P describes a circle Σ, and the sides M_1P and M_2P turn around certain fixed points I_1 and I_2 (M_1P passes through I_1 and M_2P through I_2);

(b) Q describes a circle Γ. Find the locus described by the centers of the circles Γ corresponding to different positions for the given line l_0.

53. Let l be an arbitrary line passing through the vertex A of a triangle ABC and meeting its base BC in a point M; let O_1 and O_2 be the centers of the circles circumscribed about triangles ABM and ACM. Find the locus described by the centers of the segments O_1O_2 corresponding to all possible positions of the line l.

54. A triangle ABC and a point O are given. Through O are passed three lines l_1, l_2 and l_3, such that the angles between them are equal to the angles of the triangle (the direction of the angles is taken into account); let $\bar{A}$, $\bar{B}$ and $\bar{C}$ be the points of intersection of these lines with the corresponding sides of $\triangle ABC$ (Figure 53).

(a) Prove that if O is
  1° the center of the circumscribed circle;
  2° the center of the inscribed circle;
  3° the point of intersection of the altitudes (the orthocenter)
  of triangle ABC, then O is also
  1° the orthocenter;
  2° the center of the circumscribed circle;
  3° the center of the inscribed circle of triangle $\bar{A}\bar{B}\bar{C}$.

(b) Let the point O be arbitrary and let the lines l_1, l_2 and l_3
  rotate around O. Find the locus of
  1° the centers of the circumscribed circles;
  2° the centers of the inscribed circles;
  3° the orthocenters of the triangles $\bar{A}\bar{B}\bar{C}$.

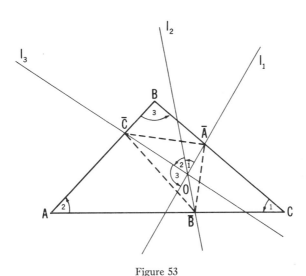

Figure 53

Let us now turn to the theorems whose proofs form the main goal of
this section.

THEOREM 3. *If the figure F moves in such a way that all positions are
similar to the original position and such that some three points A, B and
C of the figure describe three lines not passing through a common point,
then every point of the figure describes a straight line.*

THEOREM 4. *If the figure F moves in such a way that all positions are similar to the original position and so that three lines l, m and n of F, not passing through a common point, pass at all times through three given points, then every line of F passes at all times through some constant point, and every point of F describes a circle.*

PROOF OF THEOREM 3: We shall show that any two positions of F have the same center of rotation O (that is, that some point O of F remains fixed as F moves). From this it will follow that all points of F describe curves similar to the curve described by A, that is, straight lines; and this is what we are to prove.

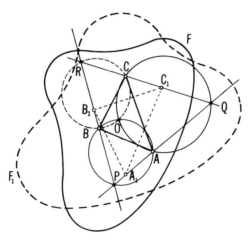

Figure 54a

Denote the points of intersection of the lines along which the points A, B and C move by the letters P, Q and R (Figure 54a).† Let F and F_1 be two positions of the figure F; let A, B, C and A_1, B_1, C_1 be the corresponding positions of the three points under consideration. The rotation center O of the figures F and F_1 is also the rotation center of the segments AB and A_1B_1, AC and A_1C_1. But, as is shown on page 44 (see Figure 31), the rotation center of the segments AB and $A'B'$ lies on the circles circumscribed about triangles ABQ and $A'B'Q$, where Q is the point of intersection of AA' and BB'; in our present case this means that O must lie on the circle cimcumscribed about $\triangle ABP$ (and on the circle circumscribed about $\triangle A_1B_1P$). In the same manner the center of rotation of the segments AC and A_1C_1 lies on the circle circumscribed about triangle ACQ (and on the circle circumscribed about $\triangle A_1C_1Q$).

† We assume that no two of the three lines are parallel. The analysis of the exceptional cases, i.e., where two lines or all three lines are parallel, is left to the reader.

Thus the rotation center of the figures F and F_1 is determined as a point of intersection of the circles circumscribed about triangles ABP and ACQ and thus *does not depend on the particular position F_1 of the figure.* But this means that any two positions of our figure have the same center of rotation, which completes the proof.

If the lines described by the points A, B and C pass through a common point P (Figure 54b), then in general the assertion of the theorem remains valid. The proof does not differ from that presented above; the common similarity center for all positions must first of all coincide with the point of intersection of the circles ABP and BCP (where A, B and C are the positions at some moment of the three points of F that we are considering), that is, with the point P. The only exception to this is the case when the circles ABP and BCP coincide, that is, the points A, B, C lie on a common circle with the point P, or what is the same thing, when

$$\sphericalangle APB + \sphericalangle ACB = 180°$$

and the points P and C lie on opposite sides of the line AB, or $\sphericalangle APB = \sphericalangle ACB$ and P and C lie on the same side of AB. In this case the assertion of the theorem need not be valid.

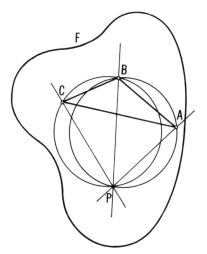

Figure 54b

PROOF OF THEOREM 4. We shall show that any two positions of F have the same rotation center O and that some point A of F describes a circle. Then it will follow by what was said on page 69 that each line of F passes at all times through some constant point (since the line l passes through a constant point) and that every point of F describes a circle (since A describes a circle).

Let us denote the points of intersections of the lines l, m and n by A, B and C† and the given points through which these lines always pass

† See footnote on p. 72.

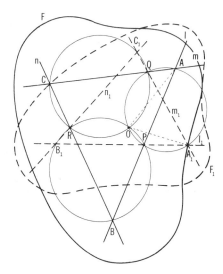

Figure 55

by P, Q and R (Figure 55). First of all it is clear that A describes a circle, since the magnitude of the angle QAP must be preserved during the motion. (The angle between the lines l and m of the figure cannot change because F remains similar to its original position.)† Further, let F and F_1 be two positions of the figure; let l, m, n and l_1, m_1, n_1 be the corresponding positions of the three lines under consideration; let A, B, C and A_1, B_1, C_1 be the corresponding positions of the points A, B and C. If O is the center of rotation for F and F_1, then angle AOA_1 is equal to the rotation angle of the spiral similarity carrying F into F_1, and, consequently, is equal to the angle between l and l_1 or to the angle between m and m_1. Thus $\angle AOA_1 = \angle APA_1 = \angle AQA_1$, that is, O lies on the circle passing through A, A_1, P and Q. In the same way one shows that O lies on the circle passing through B, B_1, P and R. From this it is clear that the rotation center O of F and F_1 is the point of intersection of the circles circumscribed about the triangles APQ and BPR and therefore *does not depend on the particular position F_1 of the moving figure*. But this means that any two positions of F have the same rotation center.

If the lines l, m and n of F all pass through a common point A, then the assertion of Theorem 4 need not be valid.

† This reasoning presupposes that the point A of the line l does not pass through the given point P (or the given point Q) during the course of its movement; in the contrary case the angle QAP is changed into its supplementary angle (see footnote ‡ on page 50 of Volume One).

55. Construct a quadrilateral $ABCD$, similar to a given one (for example, a square),

(a) whose vertices lie on four given lines;

(b) whose sides pass through four given points;

(c) whose sides BC, CD and diagonal BD pass through three given points, and whose vertex A lies on a given circle.

Problems 55 (a) and (b) can also be formulated as follows:

(a) inscribe in a given quadrilateral a quadrilateral that is similar to some other given quadrilateral (for example, a square);

(b) circumscribe about a given quadrilateral a quadrilateral that is similar to some other given quadrilateral (for example, a square).

56. Let four lines l_1, l_2, l_3, and l_4 be given. Construct a line l with the property that the three intervals cut off on it by the four given lines form given ratios.

57. Rotate each side of triangle ABC about the midpoint of that side through a fixed angle α (in the same direction each time); let $A'B'C'$ be the new triangle formed by the rotated lines. Find the locus of points of intersection of the altitudes, of the angle bisectors, and of the medians of triangle $A'B'C'$ corresponding to different values of the angle α. Prove that the centers of the circumscribed circles of all these triangles coincide.

58. Let M, K and L be three points lying on sides AB, BC and AC of triangle ABC. Prove that

(a) the circles S_1, S_2 and S_3 circumscribed about triangles LMA, MKB, and KLC meet in a point;

(b) the triangle formed by the centers of circles S_1, S_2 and S_3 is similar to triangle ABC.

The proposition contained in Problem 58 (a) can be greatly extended; see Problem 218 (b) in Section 1 of Chapter II, Volume Three. In an analogous way Problem 58 (b) can be generalized; we shall not do this, however.

59. In a circle S let there be inscribed two directly congruent (see Volume One, page 61) triangles ABC and $A_1B_1C_1$; let $\bar{A}$, $\bar{B}$, and $\bar{C}$ be the points of intersection of their corresponding sides (Figure 56). Prove that

(a) triangle $\bar{A}\bar{B}\bar{C}$ is similar to triangles ABC and $A_1B_1C_1$;

(b) the point of intersection of the altitudes of triangle $\bar{A}\bar{B}\bar{C}$ coincides with the center of the circle S.

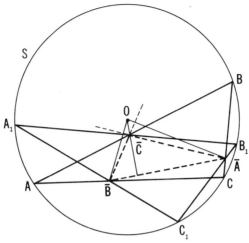

Figure 56

60. Let l be an arbitrary line in the plane and let l_1, l_2, and l_3 be three lines symmetric to l with respect to the sides of a given (non-right angled!) triangle ABC; let T be the triangle formed by the lines l_1, l_2, and l_3. Prove that

(a) all triangles T, corresponding to different positions of the initial line l, are similar to one another;

(b) all lines l such that l_1, l_2, and l_3 all meet in a point P pass through the point H of intersection of the altitudes of triangle ABC; the locus of points P of intersection of l_1, l_2, l_3 is the circumscribed circle about triangle ABC;

(c) all lines l such that the triangle T has a given area are tangent to the same circle with center H.

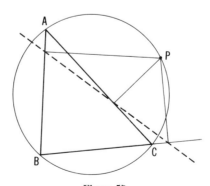

Figure 57

61. Prove that the feet of the perpendiculars dropped from a point P onto the sides of a triangle ABC all lie on a line (*Simson's line*, Figure 57) if and only if the point P lies on the circumscribed circle of $\triangle ABC$.

62. From the result of Problem 61 derive proofs of the following theorems:

(a) The four circles circumscribed about the four triangles determined by any four lines in the plane (no three of which meet in a point and no two of which are parallel) pass through a common point;

(b) Let a circle S together with three chords MA, MB and MC be given. Construct the three circles having these chords as their diameters. Each pair of these three circles intersect in another point in addition to M; prove that these points of intersection all lie on a line (Figure 58);

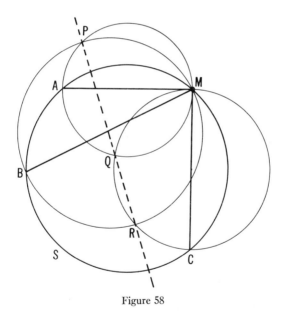

Figure 58

(c) If a, b, c, and d are the lengths of the consecutive sides of a quadrilateral $ABCD$ about which a circle can be circumscribed, and if e and f are the lengths of its diagonals, then

$$ac + bd = ef$$

(*Ptolemy's Theorem*).

In another connection Problem 62 (a) was already presented in Chapter I, Section 2 of this volume (see Problem 35 and in particular Figure 32). Ptolemy's Theorem will be presented in a different connection in Chapter II of Volume Three (see Problems 258 and 269 in Section 4); there we also present the converse to Ptolemy's Theorem (see Problem 269), and a theorem that can be regarded as a far-reaching generalization of Ptolemy's Theorem (see Problem 261 in Section 4, and Problem 273 in Section 5).

63.   Four lines are given in the plane, no three of which pass through a common point and no two of which are parallel. Prove that the points of intersection of the altitudes of the four triangles formed by these lines lie on a line.

This problem will be presented in a different connection in Section 2, Chapter I, Volume Three [see Problem 130 (b)].

64.   Find the locus of points M such that the lengths of the tangents from M to two intersecting circles S_1 and S_2 has a given ratio.

See also Problem 250 (b) in Section 3, and Problem 260 in Section 4 of Chapter II, Volume Three.

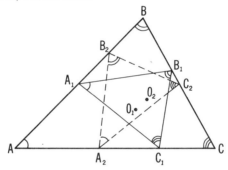

Figure 59

Inscribe in a given triangle ABC another triangle $A_1B_1C_1$ similar to ABC (the order of the letters indicates the corresponding sides) such that the vertex A_1 of triangle $A_1B_1C_1$ lies on side AB, the vertex B_1 lies on side BC and the vertex C_1 lies on side CA (Figure 59). There are infinitely many such triangles $A_1B_1C_1$ satisfying the conditions that we have posed—in an arbitrary manner one can choose the direction of one of the sides or the position of one of the vertices of triangle $A_1B_1C_1$ [see Problem 9 (b) of Section 1 and Problem 30 (a) of Section 2, Chapter I]. All such triangles $A_1B_1C_1$ may be considered images of $\triangle ABC$ under spiral similarities with the same center of rotation O_1 (see the proof of Theorem 3). The point O_1 is called the *first center of rotation* of triangle ABC. By the *second center of rotation* O_2 of triangle ABC we mean the common center of rotation of $\triangle ABC$ and the similar triangles $A_2B_2C_2$ (the order of the letters indicates the corresponding sides), where triangle $A_2B_2C_2$ is inscribed in triangle ABC in such a manner that A_2 lies on side CA, B_2 on side AB, and C_2 on side BC.

65.   Let $A'B'C'$ be a triangle similar to $\triangle ABC$ (the order of the letters indicates the order of the corresponding sides) and such that A' lies on side BC, B' on side AC, and C' on side AB of triangle ABC. Prove that the rotation center O of triangles ABC and $A'B'C'$ is the center of the circumscribed circle about triangle ABC.

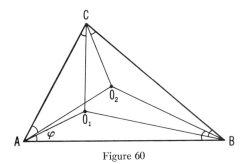

Figure 60

66.   Let O_1 and O_2 be the first and second rotation centers of triangle ABC and let O be the center of the circumscribed circle. Prove that

(a) $\angle O_1AB = \angle O_1BC = \angle O_1CA = \angle O_2BA = \angle O_2CB = \angle O_2AC$
(Figure 60); conversely, if, for example,

$$\angle MAB = \angle MBC = \angle MCA,$$

then the point M coincides with O_1;

(b) O_1 coincides with O_2 if and only if ABC is an equilateral triangle;

(c) O_1 and O_2 are equidistant from O: $O_1O = O_2O$;

(d) the common value φ of the angles O_1AB, O_1BC, O_1CA, O_2BA, O_2CB, and O_2AC, see part (a), does not exceed $30°$; $\varphi = 30°$ if and only if ABC is an equilateral triangle.

67.   Construct the rotation centers O_1 and O_2 of a given triangle ABC.

68.   Let $A_1B_1C_1$ and $A_2B_2C_2$ be two triangles inscribed in a triangle ABC and similar to it (the order of the letters indicates the order of the corresponding sides), and such that the points A_1, B_1 and C_1 lie respectively on the sides AB, BC and CA of triangle ABC, and the points A_2, B_2 and C_2 lie respectively on sides CA, AB and BC of this triangle. Suppose, in addition, that sides A_1B_1 and A_2B_2 of triangles $A_1B_1C_1$ and $A_2B_2C_2$ form equal angles with side AB of triangle ABC (Figure 61). Prove that

(a) triangles $A_1B_1C_1$ and $A_2B_2C_2$ are congruent;

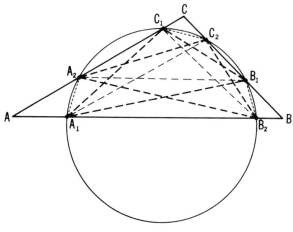

Figure 61

(b) lines B_2C_1, C_2A_1 and A_2B_1 are parallel to sides BC, CA and AB, and lines A_1A_2, B_1B_2 and C_1C_2 are anti-parallel† to these sides;

(c) the six points A_1, B_1, C_1, A_2, B_2, C_2 lie on a circle.

69. (a) Let A_1, B_1, C_1 and A_2, B_2, C_2 be the projections of the first and second rotation centers of triangle ABC onto the sides of the triangle. Prove that triangles $A_1B_1C_1$ and $A_2B_2C_2$ are both similar to triangle ABC and congruent to each other, and that the six points A_1, B_1, C_1, A_2, B_2, C_2 lie on a circle whose center is the midpoint of the segment joining the first and second rotation centers of triangle ABC (Figure 62a).

(b) Prove that in a given triangle ABC one may inscribe two triangles $A_1B_1C_1$ and $A_2B_2C_2$ such that the sides of these triangles are perpendicular to the sides of triangle ABC; moreover these two triangles are congruent to each other and the three segments joining the corresponding vertices of these two triangles are equal to each other and meet in a common point, their midpoint (Figure 62b).

70.   Let O_1 be one of the rotation centers of triangle ABC; let A', B', C' be the points of intersection of lines AO_1, BO_1, CO_1 with the circle circumscribed around triangle ABC. Prove that

† A line meeting sides AB and AC of triangle ABC in points P and Q, respectively, is said to be *anti-parallel* to side BC if $\sphericalangle APQ = \sphericalangle ACB$ and $\sphericalangle AQP = \sphericalangle ABC$ (the segment PQ is parallel to side BC if $\sphericalangle APQ = \sphericalangle ABC$ and $\sphericalangle AQP = \sphericalangle ACB$).

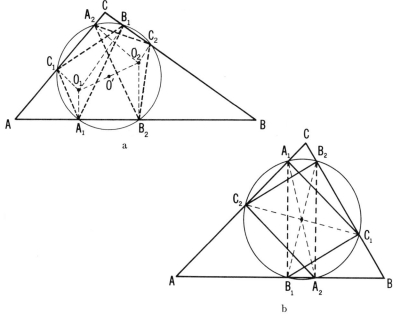

a

b

Figure 62

(a) triangle $A'B'C'$ is congruent to triangle ABC;

(b) the six triangles into which the hexagon $AC'BA'CB'$ is divided by the lines joining its vertices to the point O_1 are all similar to triangle ABC.

71. Let M be an arbitrary point inside triangle ABC. Prove that at least one of the angles MAB, MBC, MCA and at least one of the angles MAC, MCB, MCA does not exceed 30°.

To conclude this section we consider some properties of three similar figures F_1, F_2, F_3. Let O_3, O_2 and O_1 denote the centers of rotation of each successive pair of these figures. The triangle $O_1O_2O_3$ is called the *triangle of similarity*, and the circle circumscribed about this triangle is called the *circle of similarity* of the figures F_1, F_2 and F_3,† In case the

† The concept of rotation center of two directly similar figures generalizes: 1) the center of rotation of two directly congruent figures; 2) the center of similarity of two centrally similar figures. Therefore the point in question could with equal justification be called "rotation center" or "similarity center" of the figures. In the questions we consider in the conclusion of this section the second name would be more appropriate; thus we would say: the similarity centers of the successive pairs of three directly similar figures form the triangle of similarity of these figures. However the term "rotation center" is more widespread in the literature.

points O_1, O_2 and O_3 all lie on a line, or coincide, the circle of similarity degenerates to a line—the *similarity axis*—or to a point, the *similarity center*, of the figures. [If F_1, F_2 and F_3 are pairwise centrally similar then the circle of similarity degenerates to either a line or a point; see the theorem on the three centers of similarity (page 29).]

In Problems 72 and 73 we assume that the circle of similarity of the three figures F_1, F_2 and F_3 does not degenerate to a line or a point.

72.   Three similar figures F_1, F_2 and F_3 are given in the plane. Let A_1B_1, A_2B_2 and A_3B_3 be three corresponding line segments in these figures; let $D_1D_2D_3$ be the triangle whose sides are the lines A_1B_1, A_2B_2 and A_3B_3 (Figure 63). Prove that

(a) the lines D_1O_1, D_2O_2 and D_3O_3 meet in a point U that lies on the circle of similarity of the figures F_1, F_2 and F_3 (Figure 63a);

(b) the circles circumscribed about triangles $A_1A_2D_3$, $A_1A_3D_2$ and $A_2A_3D_1$ meet in a point V that lies on the circle of similarity of the figures F_1, F_2 and F_3 (Figure 63b);

(c) let $D'_1D'_2D'_3$ be some triangle different from $D_1D_2D_3$, whose sides are three corresponding lines in the figures F_1, F_2 and F_3. Then the triangles $D_1D_2D_3$ and $D'_1D'_2D'_3$ are directly similar and the rotation center O of these two triangles lies on the circle of similarity of F_1, F_2 and F_3 (Figure 63c).

73.   Let F_1, F_2 and F_3 be three similar figures; let l_1, l_2 and l_3 be corresponding lines in these figures, and suppose that l_1, l_2, l_3 meet in a point W (Figure 64). Prove that

(a) W lies on the circle of similarity of F_1, F_2 and F_3;

(b) l_1, l_2 and l_3 pass through three constant (that is, not depending on the choice of the lines l_1, l_2 and l_3) points J_1, J_2 and J_3 that lie on the circle of similarity of F_1, F_2 and F_3.

In addition to the theorems that form the content of Problems 72 and 73, one could mention many other remarkable properties possessed by three similar figures F_1, F_2 and F_3. We shall list some of these properties.

I. The triangle $J_1J_2J_3$ [see Problem 73 (b)] is oppositely similar to the triangle $D_1D_2D_3$ whose sides are any three corresponding lines of the figures F_1, F_2 and F_3.

II. The lines J_1O_1, J_2O_2, J_3O_3 meet in a single point T.

III. If three corresponding points A_1, A_2 and A_3 of F_1, F_2 and F_3 lie on a line l, then this line passes through a fixed point T (the same point described in II). Conversely, every line through T contains three corresponding points of F_1, F_2 and F_3.

IV. If A_1, A_2 and A_3 are three corresponding points of the figures F_1, F_2 and F_3, then the circles circumscribed about the triangles $A_1O_2O_3$, $A_2O_1O_3$ and $A_3O_1O_2$, meet in a point.

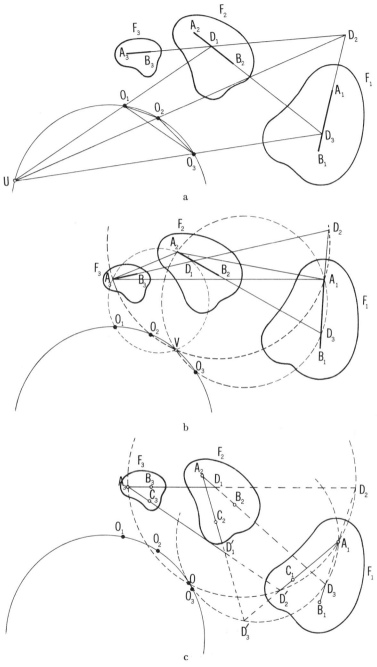

a

b

c

Figure 63

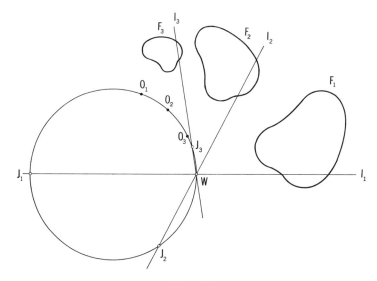

Figure 64

V. By the *basic triangle* of the point A_1 in the figure F_1 we mean the triangle $A_1A_2A_3$, where A_2 and A_3 are the points of F_2 and F_3 that correspond to the point A_1 of F_1. Then:

a) the locus of points A_1 in F_1 such that the basic triangle of A_1 has angle $A_2A_1A_3$ (or angle $A_1A_2A_3$, or angle $A_1A_3A_2$) of a given size is a circle;

b) the locus of points A_1 such that the basic triangle $A_1A_2A_3$ has side A_2A_3 (or side A_1A_3, or side A_2A_1) of a given length is a circle;

c) the locus of points A_1 such that the basic triangle $A_1A_2A_3$ has a given ratio for sides A_1A_2/A_1A_3 (or for sides A_1A_2/A_2A_3, or for sides A_1A_3/A_2A_3) is a circle;

d) the locus of points A_1 such that the basic triangle $A_1A_2A_3$ has a given area is a circle.

We leave to the reader the task of finding his own proofs for all these assertions.

2. Applications of isometries and of similarity transformations to the solution of maximum-minimum problems

The present, rather short, section is not closely connected to the rest of the book. In this section are collected a number of problems of finding the largest and smallest values of various geometric quantities. These problems are solved by various methods, in most cases by using isometries or similarity transformations; this last circumstance is the justification for including this section in the book.

74. (a) A line l and two points A, B on the same side of l are given. Find a point X on l such that the sum of the distances AX and BX has the smallest possible value.

(b) A line l and two points A and B on opposite sides of l are given. Find a point X on l such that the difference of the distances AX and BX is as large as possible.

75. (a) In a given triangle ABC inscribe another triangle, one vertex of which coincides with a given point P on side AB, and whose perimeter has the smallest possible value.

(b) Inscribe in a given triangle ABC a triangle whose perimeter has the smallest possible value.

76. Inscribe in a given quadrilateral $ABCD$ a quadrilateral of minimum perimeter. Prove that this problem does not, in general, have a proper solution (that is, a solution consisting of a non-degenerate quadrilateral). However, if the quadrilateral $ABCD$ can be inscribed in a circle, then the problem has an infinite number of solutions, that is, there are infinitely many quadrilaterals of the same perimeter inscribed in $ABCD$, such that any other inscribed quadrilateral in $ABCD$ has a larger perimeter.

One could pose the following more general problem: *in a given n-gon inscribe an n-gon of minimum perimeter*. By methods analogous to those used in solving Problems 75 (b) and 76 it can be shown that if n is *odd*, then the problem has, in general, a unique solution, whereas if n is *even*, then it either has no solution at all, or it has infinitely many solutions.

77. Inscribe in a given triangle ABC a triangle whose sides are perpendicular to the radii OA, OB, OC of the circumscribed circle. From this construction derive another solution to Problem 75 (b) for acute triangles.

78. In a given triangle ABC inscribe a triangle DEF such that the quantity $a \cdot EF + b \cdot FD + c \cdot DE$, where a, b, c are given positive numbers, has the smallest possible value.

79. In the plane of triangle ABC find a point M for which the sum of the distances to the vertices is as small as possible.

80. (a) Around a given triangle ABC circumscribe an equilateral triangle such that the perpendiculars to the sides of the equilateral triangle erected at A, B, C meet in a common point. From this construction derive another solution to Problem 79.

(b) In a given triangle ABC inscribe an equilateral triangle such that the perpendiculars dropped from points A, B, C onto the sides of the equilateral triangle meet in a point. From this construction derive still another solution to Problem 79.

81. (a) Let ABC be an equilateral triangle and let M be an arbitrary point in the plane of this triangle. Prove that $MA + MC \geq MB$. When do we have $MA + MC = MB$?

(b) From the proposition of problem (a) derive yet another solution to Problem 79.

82. Let ABC be an isosceles triangle with $AC = BC \geq AB$. For which point M in the plane of the triangle will the sum of the distances from M to A and from M to B minus the distance from M to C, that is, the quantity $MA + MB - MC$, be a minimum?

83. In the plane of a given triangle ABC find a point M such that the quantity $a \cdot MA + b \cdot MB + c \cdot MC$, where a, b, c are given positive numbers, has the smallest possible value.

It would be possible to permit some of the numbers a, b, c in Problem 83 to be negative; however in this case it would be necessary in the solution of the problem to distinguish a great many separate cases (compare the solution of Problem 82).

SOLUTIONS

Chapter One. Classification of similarity transformations

1. Suppose that the line l has been found (Figure 65). By hypothesis the point C is centrally similar to the point B with similarity center A and coefficient of similarity n/m; therefore it lies on the line l_1', centrally similar to l_1, with center A and coefficient n/m, and it can be found as the point of intersection of the lines l_2 and l_1'. If l_1 is not parallel to l_2, then the problem has a unique solution; if $l_1 \parallel l_2$, then l_1' is either parallel to l_2 or coincides with l_2, and corresponding to this the problem either has no solution at all or it is undetermined.

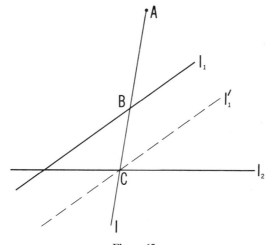

Figure 65

87

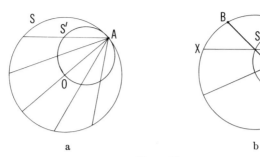

a   b

Figure 66

2. (a) The required locus is obtained from the circle S by a central similarity transformation with center A and similarity coefficient $\frac{1}{2}$; consequently, it is a circle with diameter AO, where O is the center of S (Figure 66a).

(b) Construct the circle S' having AO as diameter (O is the center of S). The points of intersection of S' with the chord BC determine the desired chord of S (Figure 66b), since S' is the locus of the midpoints of all chords of S through A [see part (a)]. The problem can have two, one, or no solutions.

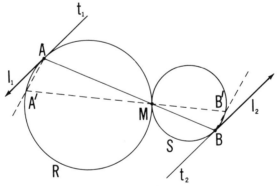

Figure 67a

3. Consider the central similarity with center M and coefficient $\mp r_2/r_1$, where r_1 and r_2 are the radii of circles R and S respectively, and the minus sign is chosen in the case of exterior tangency of the two circles (Figure 67a), while the plus sign is chosen in the case of interior tangency of the two circles (Figure 67b). This transformation carries the circle R of radius r_1 into a circle of radius r_2, tangent to R at the point M; that is, it carries R into S. The point A on the circle R is carried by this transformation into the point B on the circle S, and the tangent line t_1 to R at A is carried into the tangent line t_2 to S at B. Since the line t_2 is obtained from t_1 by a central similarity, the two lines are parallel.

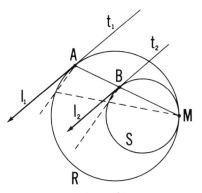

Figure 67b

4. Consider the central similarity transformation with center M and coefficient $k = r_2/r_1$, where r_1 and r_2 are the radii of circles R and S, respectively. This transformation carries the lines m and n onto themselves, and carries the circle R tangent to m and n and with radius r_1 onto a circle tangent to m and n and with radius r_2; that is, it carries R onto S (Figure 68). Also, the line l is carried into itself, the segment AB is carried into CD, the point E into F and, consequently, triangle ABE is carried onto triangle CDF. From this it follows that these triangles are similar [the assertion of part (a)], and that the similarity coefficient is $k = r_2/r_1$; therefore Area $(\triangle CDF)$: Area $(\triangle ABE) = k^2 = (r_2/r_1)^2$ [the assertion of part (b)]. Finally, from the fact that triangle CDF is obtained from triangle ABE by a central similarity with center M, it follows that the line joining two corresponding points of these triangles, for example, their centroids (the points of intersection of their medians), passes through the point M [the assertion of part (c)].

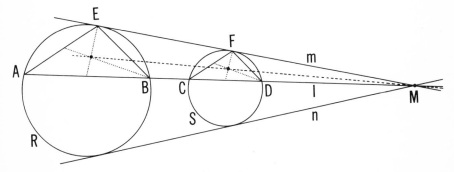

Figure 68

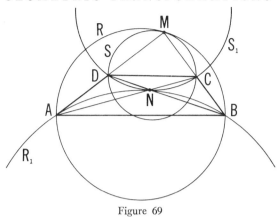

Figure 69

5. (a) The central similarity with center M and coefficient

$$k = DC/AB = MD/MA = MC/MB$$

(Figure 69) carries triangle MAB onto triangle MDC and the circle R circumscribed about triangle MAB onto the circle S circumscribed about triangle MDC. Since S is obtained from R by a central similarity whose center M lies on R, it follows that R and S are tangent at M.

(b) The central similarity with center N and coefficient

$$k_1 = CD/AB = NC/NA = ND/NB$$

(here we are considering the ratio of *directed* segments, so that k_1 is negative) carries triangle NAB onto triangle NCD, and carries the circle R_1 circumscribed about triangle NAB onto the circle S_1 circumscribed about triangle NCD. Since the similarity center N lies on R_1 the two circles are tangent to each other at N.

(c) The ratio of the radii of S and R is $k = DC/AB$ [since triangles MAB and MDC are similar; see part (a)]. The ratio of the radii of circles S_1 and R_1 is $|k_1| = |CD/AB|$ [see the solution to part (b)]. But clearly $|CD/AB| = DC/AB$, which is the assertion of part (c).

6. (a) If M is the point of intersection of the two non-parallel sides AD and BC of the trapezoid (Figure 70a), then the central similarity with center M and coefficient DC/AB carries the segment AB onto DC and carries $\triangle ABE$ into $\triangle DCF$ [compare the solution to Problem 5 (a)]. The assertion of the problem follows from this.

(b) If N is the point of intersection of the diagonals AC and BD of the trapezoid $ABCD$ (Figure 70b), then the central similarity with center N and (negative!) coefficient CD/AB carries the segment AB into CD and carries one of the squares into the other [compare the solution to Problem 5 (b)]. The assertion of the problem follows from this.

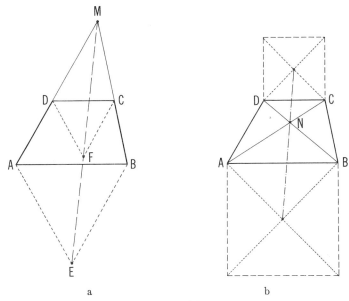

a   b

Figure 70

7. The central similarity with center at the point M of intersection of the extensions of sides AD and BC of trapezoid $ABCD$ and with coefficient DC/AB carries the segment AB onto the segment DC and carries the midpoint K of AB into the midpoint L of side DC [compare the solution to Problem 6 (a)]. Therefore, the line KL passes through the similarity center M (Figure 71). The point K is also carried onto the point L by the central similarity with center at the point N of intersection of the diagonals AC and BD of the trapezoid and with the (negative) coefficient CD/AB. This transformation carries the segment AB onto CD [compare the solution to Problem 6 (b)]. Therefore the line KL also passes through the point N.

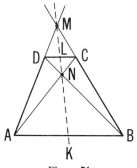

Figure 71

8. (a) Suppose that the required line l has been drawn, so that $AB/AC = 1/2$ (Figure 72a). Then we are led to the following solution. Construct a circle S_2', centrally similar to S_2 with center of similarity at an arbitrary point A of the circle S_1 and with similarity coefficient $\frac{1}{2}$. The point A and a point of intersection of the circles S_2 and S_2' determine the desired line. (Once A is selected there may be two such lines, or exactly one line, or no line at all.)

(b) Again we are led to the following construction by first supposing that l has been drawn so that $AB/AC = 1/2$ (Figure 72b). Construct a circle S_3' centrally similar to S_3 with similarity center at an arbitrary point A of the circle S_1 and with similarity coefficient $\frac{1}{2}$. The point A and a point of intersection of the circles S_2 and S_3' determine the desired line. (Once A is selected there may be two, one, or no such lines.)

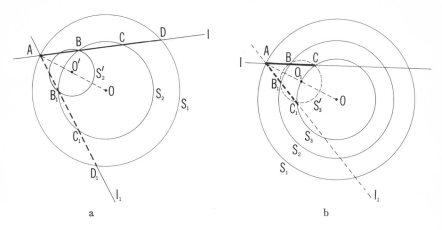

a　　　　　　　　b

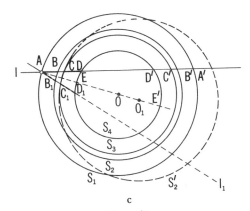

c

Figure 72

(c) Suppose that the line l has been found, and let A', B', C' and D' be the four other points of intersection of it with the circles S_1, S_2, S_3 and S_4 (Figure 72c). Clearly $AB = D'C'$. Hence

$$AD' = AB + BC' - D'C' = BC'$$

and therefore,

$$\frac{AD}{BC} = \frac{AD \cdot AD'}{BC \cdot BC'}.$$

The quantity $AD \cdot AD'$ does not depend on the position of the points D and D' where a line from A intersects S_4; it is equal to $AE \cdot AE'$, where E and E' are the points of intersection of line AO (O is the common center of all four circles) with the circle S_4, that is, it is equal to $(r_1 - r_4) \cdot (r_1 + r_4)$, where r_1 and r_4 are the radii of S_1 and S_4. Analogously $BC \cdot BC' = (r_2 - r_3) \cdot (r_2 + r_3)$, where r_2 and r_3 are the radii of S_2 and S_3. Thus

$$\frac{AD}{BC} = \frac{r_1^2 - r_4^2}{r_2^2 - r_3^2}.$$

Moreover,

$$AD = AB + BC + CD = 2AB + BC;$$

consequently,

$$\frac{AD}{BC} = 2\frac{AB}{BC} + 1.$$

Hence

$$\frac{AB}{BC} = \frac{1}{2}\left(\frac{r_1^2 - r_4^2}{r_2^2 - r_3^2} - 1\right) = \frac{r_1^2 - r_2^2 + r_3^2 - r_4^2}{2(r_2^2 - r_3^2)}$$

and thus

$$\frac{AC}{AB} = \frac{AB + BC}{AB} = 1 + \frac{BC}{AB} = 1 + \frac{1}{AB/BC} = \frac{r_1^2 + r_2^2 - r_3^2 - r_4^2}{r_1^2 - r_2^2 + r_3^2 - r_4^2},$$

so that the quantity AC/AB can be regarded as known.

From this we have the following construction. Perform a central similarity with center at an arbitrary point A on the circle S_1 and with coefficient

$$k = \frac{r_1^2 + r_2^2 - r_3^2 - r_4^2}{r_1^2 - r_2^2 + r_3^2 - r_4^2}.$$

Let the circle S_2 be carried by this transformation onto a circle S_2'. The point A and a point of intersection of the circles S_2' and S_3 determine the desired line (once A is selected there may be two, one, or no lines).

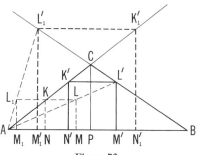

Figure 73

9. (a) Construct a square $KLMN$ such that K lies on side AC and MN lies on the base AB (Figure 73); if L' is the point of intersection of the line AL with BC, then the central similarity with center A and coefficient $k = AL'/AL$ carries $KLMN$ into the desired square $K'L'M'N'$.

[If we insist that the desired square have all its vertices on the actual sides of triangle ABC, and *not on their extensions*, then clearly the problem will have a unique solution in case both angles A and B are less than or equal to 90°, and will have no solutions at all if either of these angles is greater than 90°. If we permit the vertices of the square to lie *on the extensions* of the sides of triangle ABC [see the footnote to the statement of Problem 10 (c)], then, in general, the problem will have two solutions, represented by the two squares $K'L'M'N'$ and $K'_1L'_1M'_1N'_1$ in Figure 73. Only in the case when $AL_1 \parallel BC$ (in the notations of Figure 73), will the problem have a unique solution. Now take $K = C$ in Figure 73 so that the altitude CP of the triangle is a common side of the two squares $KLMN$ and KL_1M_1N, namely the side KN. Then we can easily convince ourselves that the problem has a unique solution, that is, $AL_1 \parallel BC$, if and only if the *altitude CP of triangle ABC is equal to the base AB.*†]

(b) Construct a triangle LMN whose sides are parallel to l_1, l_2, and l_3, and such that L lies on BC and M on CA. If N' is the point of intersection of the lines CN and AB, then the central similarity with center C and coefficient $k = CN'/CN$ carries triangle LMN onto the desired triangle $L'M'N'$ (Figure 74).

† This result can also be derived from the following elegant theorem, whose proof we leave to the reader:

If $K'L' = m$ and $K_1'L_1' = n$ are sides of the two squares $K'L'M'N'$ and $K'_1L'_1M'_1N'_1$ inscribed in triangle ABC in accordance with the conditions of Problem 9 (a), and if $CP = h$ is the altitude of triangle ABC, then the segment h is the harmonic mean of m and n, that is

$$\frac{1}{h} = \frac{1}{2}\left(\frac{1}{m} + \frac{1}{n}\right), \quad \text{or} \quad h = \frac{2mn}{m+n}.$$

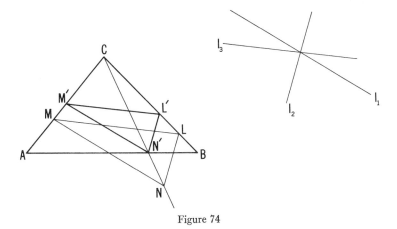

Figure 74

10. (a) Pass lines through A and B parallel to l_3 and l_4, respectively, and through their intersection point O draw lines l_1' and l_2' parallel to l_1 and l_2 (Figure 75). Choose two pairs of points M_1, M_2 and N_1, N_2 that satisfy the hypotheses of the problem: M_1 and N_1 lie on l_1, M_2 and N_2 are on l_2, and $AM_1/BM_2 = AN_1/BN_2 = m$.

Let M_1', N_1' be the points of intersection of the lines through the points M_1, N_1 and parallel to l_3 with the line l_1'. Let M_2', N_2' be the points of intersection of the lines through M_2, N_2 and parallel to l_4 with the line l_2'. Then

$$\frac{OM_1'}{OM_2'} = \frac{ON_1'}{ON_2'} \ (= m), \qquad \text{or} \qquad \frac{OM_1'}{ON_1'} = \frac{OM_2'}{ON_2'}.$$

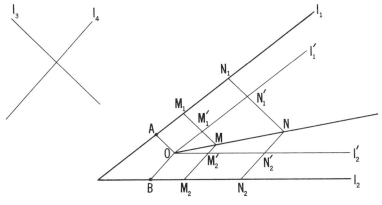

Figure 75

It follows that the pair of lines M_1M_1' and M_2M_2' are centrally similar to the pair of lines N_1N_1' and N_2N_2' with similarity center O. Consequently, the point M of intersection of the lines M_1M_1' and M_2M_2' is centrally similar with similarity center O to the point N of intersection of the lines N_1N_1' and N_2N_2'; in other words, any two points of the desired locus lie on one and the same line through O. Hence the required locus is a line; to construct it, it is enough to observe that it passes through O and through an arbitrary point M satisfying the conditions of the problem.

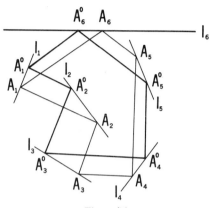

Figure 76

(b) Let $A_1^{\circ}A_2^{\circ} \cdots A_n^{\circ}$ be some fixed position of our polygon (see Figure 76, where $n = 6$). Since the sides of the original polygon are always parallel to the corresponding sides of $A_1^{\circ}A_2^{\circ} \cdots A_n^{\circ}$, and the vertices $A_1, A_2, \cdots, A_{n-1}$ slide along the lines $l_1, l_2, \cdots, l_{n-1}$, it follows that the ratios

$$\frac{A_1^{\circ}A_1}{A_2^{\circ}A_2}, \quad \frac{A_2^{\circ}A_2}{A_3^{\circ}A_3}, \quad \cdots, \quad \frac{A_{n-2}^{\circ}A_{n-2}}{A_{n-1}^{\circ}A_{n-1}}$$

remain constant; therefore the ratio

$$\frac{A_1^{\circ}A_1}{A_{n-1}^{\circ}A_{n-1}} = \frac{A_1^{\circ}A_1}{A_2^{\circ}A_2} \cdot \frac{A_2^{\circ}A_2}{A_3^{\circ}A_3} \cdot \cdots \cdot \frac{A_{n-2}^{\circ}A_{n-2}}{A_{n-1}^{\circ}A_{n-1}}$$

also remains constant. By part (a) this means that vertex A_n also moves on a line l_n (determined by any two positions of this vertex).

(c) Let $l_1, l_2, \cdots, l_n$ be the lines on which the sides of the given polygon lie. Choose the point A_1 arbitrarily on the line l_1, and construct a polygon $A_1A_2 \cdots A_n$ whose sides are parallel to the given lines, and whose vertices $A_2, \cdots, A_{n-1}$ lie on the lines $l_2, \cdots, l_{n-1}$. If the vertices $A_1, A_2, \cdots, A_{n-1}$ of this polygon slide along the lines $l_1, l_2, \cdots, l_{n-1}$ in such a manner that the sides remain parallel to the given lines, then, by part (b), the vertex A_n also slides along some line m, determined by any two positions of the vertex A_n. Then the intersection of the line m with the line l_n determines the position of the vertex $\bar{A}_n$ of the desired polygon $\bar{A}_1\bar{A}_2\bar{A}_3 \cdots \bar{A}_n$.

If l_n is not parallel to m, then the problem has a unique solution; if $l_n \parallel m$, but $l_n \neq m$, then the problem has no solution; if $l_n = m$, then the problem is undetermined.

11. Since side BC does not change in length and B is fixed, C moves on a circle with center B as the "hinged" parallelogram moves. But Q is obtained from C by a central similarity transformation with center at the fixed point A and with coefficient $\frac{1}{2}$. Therefore Q moves on the circle obtained by this transformation from the circle on which C moves.

12. (a) Let r be the radius of the inscribed circle S of triangle ABC, and let ρ be the radius of the escribed circle σ (Figure 77). The central similarity with center A and coefficient r/ρ carries σ onto S; the line BC tangent to σ at E is carried onto a line $B'C'$ tangent to S at D_1. Since D_1 is obtained from E by a central similarity with center A, the line ED_1 passes through A. Since the two tangent lines BC and $B'C'$ to S are parallel, their points D and D_1 of tangency with S must be diametrically opposite. This proves the assertion of the problem.

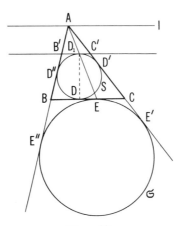

Figure 77

(b) Let the inscribed circle S touch the sides BC, CA and AB of triangle ABC in the points D, D' and D'' respectively, and let the escribed circle σ touch these sides (extended) in the points E, E' and E'' respectively (here we are speaking of the escribed circle that touches side BC and the extensions of the other two sides—see Figure 77). Let us denote the sides of triangle ABC by $BC = a$, $CA = b$, $AB = c$. The two tangent segments drawn from a point outside a circle to the circle have the same length. Consequently,

$$CD + CD' = (a - BD) + (b - AD')$$
$$= a - BD'' + b - AD''$$
$$= a + b - AB = a + b - c.$$

Since $CD = CD'$, we have

$$CD = \tfrac{1}{2}(a + b - c).$$

Analogously we have

$$AE' + AE'' = (b + CE') + (c + EE'')$$
$$= b + CE + c + BE$$
$$= b + c + BC = a + b + c,$$

and since $AE' = AE''$,

$$AE' = \tfrac{1}{2}(a + b + c).$$

Also,

$$CE = CE' = AE' - AC = \tfrac{1}{2}(a + b + c) - b = \tfrac{1}{2}(a - b + c).$$

From this it follows that

$$ED = CD - CE = \tfrac{1}{2}(a + b - c) - \tfrac{1}{2}(a - b + c) = b - c.$$

To construct triangle ABC from the given quantities r, h and $b - c$ we first construct the right triangle EDD_1 (Figure 77) with the two perpendicular sides $ED = b - c$ and $DD_1 = 2r$. Then we construct the circle S having DD_1 as diameter. At a distance h from the line ED we pass a parallel line l. The line ED_1 intersects l in the vertex A of the desired triangle. The two remaining vertices B and C are found as the points of intersection of the line ED with the tangent lines AD'' and AD' from the point A to the circle S.

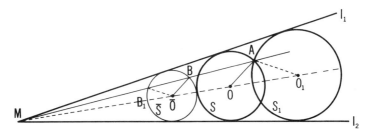

Figure 78a

13. (a) If l_1 and l_2 are parallel, then the problem has a simple solution. If l_1, l_2 are not parallel, let M be their point of intersection and consider the angular region $l_1 M l_2$ containing A. Inscribe an arbitrary circle $\bar{S}$ in the angle $l_1 M l_2$ (Figure 78a). Let B denote the point of intersection of the line MA with the circle $\bar{S}$. The desired circle S will be centrally similar to $\bar{S}$ with similarity center M and similarity coefficient MA/MB; knowing the center of similarity and the similarity coefficient, one can easily construct S. If A does not lie on l_1 or on l_2, then the problem has two solutions.

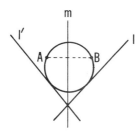

Figure 78b

(b) Let m be perpendicular to AB at its midpoint; let l' be a line symmetric to l with respect to m (Figure 78b). The desired circle S must be tangent to the line l'. Thus we are reduced to part (a): construct a circle tangent to two given lines l and l' and passing through a given point A.

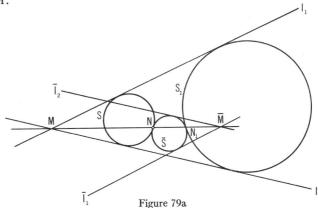

Figure 79a

(c) Denote by M the point of intersection of lines l_1 and l_2, and let S be the desired circle (Figure 79a).† The point N of tangency of the

† If $l_1 \parallel l_2$, the problem can be solved easily, since then the radius of S is immediately determined.

circles $\bar{S}$ and S is their similarity center. The central similarity carrying S into $\bar{S}$ carries the lines l_1 and l_2 into lines $\bar{l}_1 \parallel l_1$ and $\bar{l}_2 \parallel l_2$, tangent to $\bar{S}$; the point M is carried into the point $\bar{M}$ of intersection of $\bar{l}_1$ and $\bar{l}_2$. Hence N can be found as the point of intersection of the line $M\bar{M}$ with the circle $\bar{S}$; after this S can be constructed easily (the central similarity with center N carrying $\bar{l}_1$ into l_1 also carries $\bar{S}$ into S).

Each of the lines $\bar{l}_1$ and $\bar{l}_2$ can be chosen in two ways; the line joining M to their point of intersection can meet the circle $\bar{S}$ in two points. Thus the problem can have up to eight solutions (see, for example, Figure 79b).

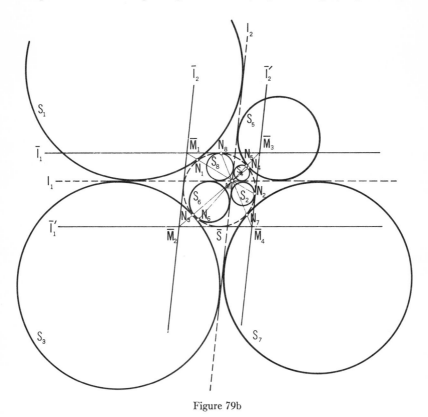

Figure 79b

14. (a) By a well-known property of medians, the triangle $A'B'C'$ formed by the lines joining the midpoints of the sides of triangle ABC is centrally similar to ABC with similarity center M (the centroid of $\triangle ABC$) and similarity coefficient $-\frac{1}{2}$ (Figure 80a). The altitudes AD, BE and CF of $\triangle ABC$ will correspond to altitudes $A'D'$, $B'E'$ and $C'F'$ of $\triangle A'B'C'$, and the point of intersection of the altitudes (the ortho-center) of $\triangle ABC$ will correspond to the point of intersection of the altitudes of $\triangle A'B'C'$, that is, the point O (since the altitudes of $\triangle A'B'C'$ are perpendicular to the sides of $\triangle ABC$ at their midpoints). Hence points H and O are centrally similar with similarity center M and coefficient $-\frac{1}{2}$, that is, they lie on the same line through M, and M divides the segment HO in the ratio $HM/MO = 2$.

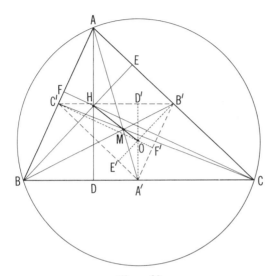

Figure 80a

(b) The proof is similar to the solution of part (a) and is based on the fact that the lines through the midpoints of the sides of the triangle and parallel to the angle bisectors are centrally similar to the angle bisectors of the triangle with similarity center M and similarity coefficient $-\frac{1}{2}$ (Figure 80b); therefore they meet in a common point K (which is the image of the point Z of intersection of the angle bisectors).

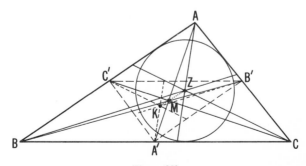

Figure 80b

(c) Let A', B', C' be the midpoints of the sides of triangle ABC; let K, K_1, K_2 and P, Q, R be the points of tangency of the sides with the escribed circle in angle A and with the inscribed circle, respectively (Figure 81). Let us prove that $AK \parallel ZA'$. Denote the sides of triangle ABC by a, b and c, the perimeter by $2p$, the altitude $A\bar{P}$ dropped on side BC by h_a, the radius of the inscribed circle by r, and the area of the triangle by S. Since $ah_a = 2pr(= 2S)$, we have $h_a/r = 2p/a$. Therefore,

$$\frac{A\bar{P}}{ZP} = \frac{2p}{a} \; ;$$

We shall show that the ratio $K\bar{P}/A'P$ is equal to this quantity.

Indeed,

$$BP = \frac{a^2 + c^2 - b^2}{2a} \; ,$$

since $b^2 = a^2 + c^2 - 2aB\bar{P}$; on the other hand,

$$BK = BK_1 = AK_1 - AB = p - c = \frac{a + b - c}{2} \; ,$$

since

$$AK_1 = \tfrac{1}{2}(AK_1 + AK_2) = \tfrac{1}{2}(AB + BK_1 + AC + CK_2)$$

$$= \tfrac{1}{2}(AB + BK + AC + CK) = \tfrac{1}{2}(a + b + c) = p.$$

Consequently, for example, in the case pictured in Figure 81,

$$K\bar{P} = B\bar{P} - BK = \frac{a^2 + c^2 - b^2}{2a} - \frac{a + b - c}{2}$$

$$= \frac{a^2 + c^2 - b^2 - a^2 - ab + ac}{2a} = \frac{(c - b)(c + b + a)}{2a} = \frac{p(c - b)}{a} \; .$$

Further

$$BP = \tfrac{1}{2}(BP + BQ) = \tfrac{1}{2}(BC - CP + BA - AQ)$$

$$= \tfrac{1}{2}(BC + BA - CR - AR) = \tfrac{1}{2}(a + c - b) = p - b$$

and, consequently,

$$A'P = BP - BA' = p - b - \frac{a}{2} = \frac{c - b}{2} \; .$$

Thus we have:

$$\frac{K\bar{P}}{A'P} = \frac{p(c - b)}{a} \div \frac{c - b}{2} = \frac{2p}{a} = \frac{A\bar{P}}{ZP} \; .$$

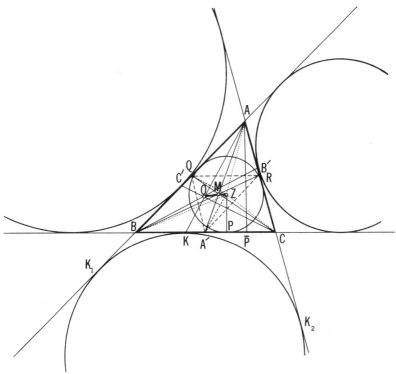

Figure 81

Hence triangles $A\bar{P}K$ and ZPA' are similar, and therefore lines AK and ZA' are parallel.

Further, in a manner entirely analogous to the solution of the previous exercises, it can be shown that the three lines through the vertices of the triangle, parallel to lines $A'Z$, $B'Z$ and $C'Z$ (these are, as we have shown, the lines joining the vertices of the triangle with the points of tangency of the opposite sides with the corresponding escribed circles), meet in a point J, centrally similar to Z with similarity center M and coefficient -2.

15. Let O be the center of the circle S circumscribed about the desired triangle ABC (Figure 82). By the result of Problem 14 (a), the centroid M of this triangle lies on the segment OH and divides this segment in the ratio $OM:MH = 1:2$; thus M can be constructed. Further, the point M divides the median AD of triangle ABC in the ratio $AM:MD = 2:1$; therefore we can construct the midpoint D of side BC. Now join the point D to the center O of circle S; since OD bisects the desired chord BC of circle S with center O, it follows that $OD \perp BC$. Therefore we can construct chord $BC \perp OD$, which completes the construction of triangle ABC.

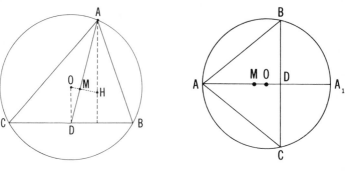

Figure 82   Figure 83

16. First note that the *point of intersection of the medians of triangle ABC, inscribed in circle S,* lies inside $\triangle ABC$ and therefore *lies inside circle S* also. On the other hand, *for any point M inside S we can inscribe a triangle ABC in S, having M as its centroid.* For this it is sufficient to choose a point D on the diameter AA_1 through M (where $AM \le MA_1$), such that $AM:MD = 2:1$ (Figure 83) and then to pass a chord BC of the circle through D such that $BC \perp OD$. *Thus the locus of centroids of all triangles inscribed in circle S is exactly the set k of all points that lie inside S* (Figure 84a). By the result of Problem 14 (a), if H is the orthocenter of $\triangle ABC$ inscribed in the circle S with center O, and if M is the centroid of this triangle, then $OH:OM = 3:1$. It follows from this that *the locus of orthocenters of all triangles inscribed in circle S coincides with the set K of all points that lie inside the circle Σ, concentric with S and having a radius three times as large as the radius of S* (Figure 84b).

It only remains to point out that the orthocenter of an acute-angled triangle ABC lies inside the triangle, and therefore lies inside S (Figure 85a). The orthocenter of a right angled triangle ABC inscribed in S with right angle at the vertex A coincides with A, and therefore lies on S (Figure 85b). Finally, the orthocenter of an obtuse-angled triangle ABC inscribed in S with obtuse angle at A lies past A on the extension of the altitude AP, and therefore lies outside of S (Figure 85c). Therefore, *the locus of orthocenters of all acute triangles inscribed in S coincides with the set k of all points that lie inside S; the locus for right triangles inscribed in S coincides with the circle S; the locus for obtuse triangles inscribed in S consists of the points in the ring between the two concentric circles S and Σ* (Figure 84b).

Finally, since the centroid of a triangle ABC inscribed in S is obtained from the orthocenter by means of a central similarity with center at the center O of S and with coefficient $1/3$, it follows that *the locus of centroids of all acute triangles inscribed in S coincides with the set κ of all points that lie inside the circle σ, concentric with S, and having radius equal to 1/3 the radius of S. In the same way the locus of centroids of all right triangles inscribed in S coincides with the circle σ, and the locus of centroids of all obtuse triangles inscribed in S coincides with the set of points in the ring between σ and S* (Figure 84a).

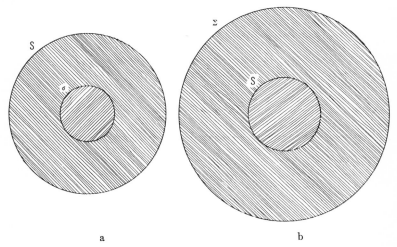

a   b

Figure 84

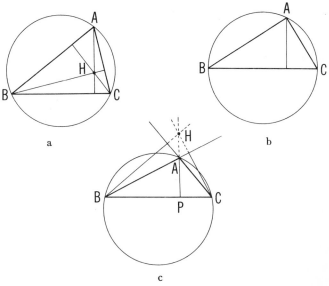

a   b

c

Figure 85

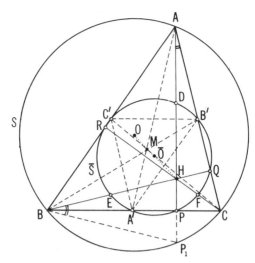

Figure 86a

17. (a) Let AP, BQ, CR denote the altitudes of triangle ABC; let A', B', C' denote the midpoints of its sides; let D, E, F denote the midpoints of the segments HA, HB and HC (Figure 86a). Clearly D, E and F lie on the circle $\bar{S}$. Let us now show that the points P, Q and R also lie on this circle.

Extend the altitude AP until it intersects S at point P_1, and join P_1 to the point B. In the right triangles APC and BQC,

$$\angle CBQ = \angle CAP \ (= 90° - \angle BCA).$$

Further, $\angle CAP = \angle CBP_1$ (since they subtend the same arc). Consequently, $\angle HBP = \angle PBP_1$, so that triangle HBP_1 is isosceles and $HP = \frac{1}{2}HP_1$; therefore the point P_1 is taken into the point P by the central similarity with center H and coefficient $\frac{1}{2}$. In exactly the same way one shows that the points Q and R are centrally similar to some points of the circumscribed circle, with similarity center H and coefficient $\frac{1}{2}$.

Further, it is clear that the points A', B' and C' lie on the circle S' which is centrally similar to S with similarity center M and coefficient $-\frac{1}{2}$. Let us show that S' coincides with $\bar{S}$. The radius of $\bar{S}$ is equal to $r/2$ (where r is the radius of S), and its center $\bar{O}$ is at the midpoint of the segment HO (O is the center of S) (Figure 86a). The radius of the circle S' is also equal to $r/2$, and the center O' of this circle lies on the line MO, while $MO'/MO = -\frac{1}{2}$. By the result of Problem 14 (a) the points O, H and M lie on a line, and $HM = 2MO$; it follows at once that the points $\bar{O}$ and O' coincide, that is, the circles $\bar{S}$ and S' coincide, which was to be proved.

We note also that A' and D, B' and E, C' and F are diametrically opposite points on the Euler circle (since, for example, the angle $A'PD$ inscribed in this circle and subtending the chord $A'D$ is a right angle). Therefore triangles $A'B'C'$ and DEF are symmetric with respect to the center of this circle and, consequently, are congruent.

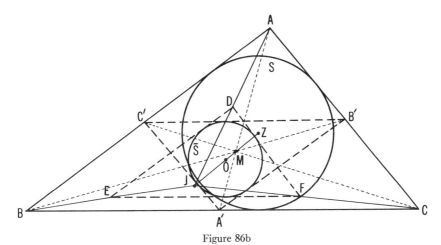

Figure 86b

(b) First of all it is evident that triangle DEF is centrally similar to triangle ABC with similarity center J and coefficient $\frac{1}{2}$ (Figure 86b). It follows that the circle $\bar{S}$, inscribed in triangle DEF, is centrally similar to the circle S inscribed in triangle ABC, with similarity center J and coefficient $\frac{1}{2}$. Further, the circle S', inscribed in triangle $A'B'C'$, is centrally similar to the circle S, with similarity center M and coefficient $-\frac{1}{2}$. But by the result of Problem 14 (c) the midpoint O of segment JZ (the center of $\bar{S}$) coincides with the center O' of circle S', which lies on the line MZ with $MO'/MZ = -\frac{1}{2}$ (compare the solution of part (a)). Therefore the circles $\bar{S}$ and S' coincide, which was to be proved.

Note also that the lines $A'B'$ and DE, $B'C'$ and EF, $C'A'$ and FD are tangent to $\bar{S}$ at diametrically opposite points (because $A'B' \| AB \| DE$, $B'C' \| BC \| EF$, $C'A' \| CA \| FD$|, and that therefore triangles DEF and $A'B'C'$ are symmetric with respect to the center of $\bar{S}$ and therefore congruent.

18. (a) Let $A_1A_2A_3A_4$ be an arbitrary quadrilateral, and let M_4, M_3, M_2 and M_1 be the centroids of the triangles $A_1A_2A_3$, $A_1A_2A_4$, $A_1A_3A_4$ and $A_2A_3A_4$ (Figure 87a). It is enough to prove that any two of the segments A_1M_1, A_2M_2, A_3M_3 and A_4M_4 are divided by their point of intersection in the ratio $3:1$; for, then it would follow that all these segments meet in a single point. Let N be the point of intersection of A_1M_1 and A_2M_2, and let L be the midpoint of A_3A_4. Then

$$A_1M_2/M_2L = A_2M_1/M_1L = 2/1;$$

consequently, triangles A_1LA_2 and M_2LM_1 are similar, and $M_2M_1 \parallel A_1A_2$, $A_1A_2/M_2M_1 = 3/1$. And now we see that triangles A_1NA_2 and M_1NM_2 are similar; therefore $A_1N/NM_1 = A_2N/NM_2 = 3/1$, which was to be proved.

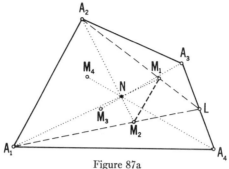

Figure 87a

In a similar manner the theorem may be proved for an arbitrary polygon. For example, if N_1 and N_2 are the centroids of the quadrilaterals $A_2A_3A_4A_5$ and $A_1A_3A_4A_5$, if M is the centroid of triangle $A_3A_4A_5$, and if P is the point of intersection of A_1N_1 and A_2N_2 (Figure 87b), then $A_1N_2/N_2M = A_2N_1/N_1M = 3/1$; therefore triangles A_1MA_2 and N_2MN_1 are similar, so $N_2N_1 \parallel A_1A_2$, and $A_1A_2/N_2N_1 = 4/1$. From this it follows that triangles A_1PA_2 and N_1PN_2 are similar, and

$$A_1P/PN_1 = A_2P/PN_2 = 4/1.$$

Thus any two segments joining the vertices of a pentagon with the centroids of the quadrilaterals formed by the four remaining vertices are divided by the point of intersection in the ratio 4:1, from which the assertion immediately follows for pentagons.

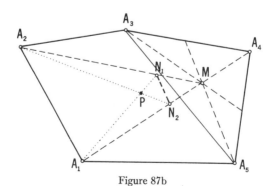

Figure 87b

(b), (c) It follows from (a) that the centroids of the four triangles whose vertices are the vertices of a quadrilateral inscribed in a circle S lie on a circle S' of radius $R/3$; from the result of Problem 17 (a) it follows that the centers of the Euler circles of these triangles are centrally symmetric to the centroids, with similarity center at the center O of the circle S, and with similarity coefficient $3/2$; therefore the centers of the Euler circles lie on a circle $\bar{S}$ of radius $(R/3)(3/2) = R/2$, from which follows the assertion of (b) for quadrilaterals. Further, if N is the centroid of the quadrilateral, O' is the center of circle S' and $\bar{O}$ is the center of $\bar{S}$, then O, N and O' lie on a line and $ON:NO' = 3:1$, O, O' and $\bar{O}$ lie on a line and $O\bar{O}:OO' = 3:2$. Therefore, O, N and $\bar{O}$ lie on a line and $ON:N\bar{O} = 2:2$, which was required to establish (c) for quadrilaterals. Now by exactly the same reasoning we may establish the assertions of (b) and (c) for pentagons, etc.; we leave it to the reader to carry this out for himself.

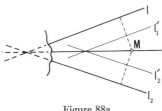

Figure 88a

19. (a) Let us perform a central similarity with center at the point M and a small enough similarity coefficient k so that the lines l_1' and l_2', into which l_1 and l_2 are taken, intersect within the confines of our picture (Figure 88a). The line joining the point M with the point of intersection of l_1' and l_2' is the desired line.

The problem is solved in the same manner in case the inaccessible point is given not by two lines, but by a line and a circle or by two circles (intersecting outside the picture).

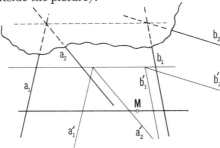

Figure 88b

(b) Carry out a central similarity with center at the point M and with coefficient k sufficiently small so that the lines a_1, a_2 and b_1, b_2 are carried into lines a_1', a_2' and b_1', b_2' such that the line joining the points of intersection of a_1' and a_2', b_1' and b_2' lies within the confines of the picture (Figure 88b). The line through M parallel to this line is the desired line.

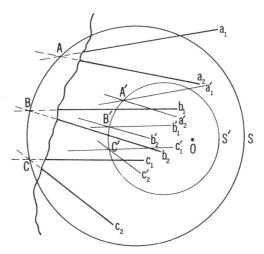

Figure 88c

(c) Carry out a central similarity with center at any point O that is accessible to us and with sufficiently small coefficient k so that the points A, B, C are carried into points A', B', C' lying within the confines of our picture (Figure 88c). Let S' be the circle circumscribed about triangle $A'B'C'$. The desired circle S is centrally similar to S' with similarity center O and coefficient $1/k$; this enables us to find its center and radius.

20. Suppose that we are required to carry out some construction on a bounded portion κ of the plane, and suppose that the diagram T, required for the completion of this construction, is not contained entirely within κ. Carry out a central similarity with center at some point O in κ and with coefficient k sufficiently small so that the transformed figure T' lies entirely in κ. Then T' can be constructed. After this T may be regarded as finished since it is centrally similar to the already completed figure T' with similarity center O and coefficient $1/k$. (If some point A of T is centrally similar to the point A' of T', and if A does not lie on our given portion of the plane, then it can be defined by a pair of lines l_1 and l_2, centrally similar to some pair of lines l_1' and l_2' passing through A'; also l_1' and l_2' can always be chosen so that l_1 and l_2 pass across a part of our picture: for this it is sufficient, for example, for the distance of the lines l_1' and l_2' from the similarity center O to be sufficiently small.)

21. This proposition is a special case of the theorem on three centers of similarity (see page 29).

22. Let S be the desired circle; construct an arbitrary circle S_1, tangent to l_1 and l_2 (Figure 89; we do not consider in the figure the situation when

$l_1 \parallel l_2$, in which case the solution to the problem is much simpler). The similarity center of circles S and S_1 is the point M of intersection of l_1 and l_2; the similarity center of circles S_1 and $\bar{S}$ can be constructed; the similarity center of circles $\bar{S}$ and S is their point of tangency N. By the theorem on three centers of similarity (see page 29) the points M, O and N lie on a line, that is, N is a point of intersection of the line OM with the circle $\bar{S}$. Having found N we can construct the circle S without difficulty.

The similarity center O of circles $\bar{S}$ and S_1 can be chosen in two ways; each of the two lines OM obtained in this manner can intersect $\bar{S}$ in two points; thus up to four circles can be constructed, all satisfying the conditions of the problem; the centers of these circles lie on one or the other of the angle bisectors of the two adjacent angles formed by the lines l_1 and l_2 (the center of circle S_1 lies on this bisector). Choosing then for S_1 the circle with center on the bisector of the other angle, we can find up to four more solutions. Thus in all the problem can have as many as eight solutions.

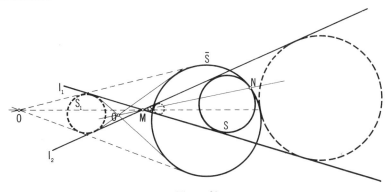

Figure 89

23. We consider the general case of n circles. Let S_1 and S_2 be tangent (externally or internally) at M_1, let S_2 and S_3 be tangent at M_2, $\cdots$, let S_{n-1} and S_n be tangent at M_{n-1}, and let S_n and S_1 be tangent at M_n. Let A_1 be an arbitrary point of S_1, and let A_2 be the second point of intersection of the line A_1M_1 with S_2, let A_3 be the second point of intersection of A_2M_2 with S_3, $\cdots$, let A_n be the second point of intersection of $A_{n-1}M_{n-1}$ with S_n, and let A_{n+1} be the second point of intersection of A_nM_n with S_1 (compare Figure 21a, b). Let r_1, r_2, $\cdots$, r_{n-1}, r_n denote the radii of the circles S_1, S_2, $\cdots$, S_{n-1}, S_n, respectively. The central similarity transformation with center M_1 and coefficient $k_1 = \mp r_2/r_1$, where the minus sign is chosen in case S_1 and S_2 are externally tangent, while the plus sign is chosen in the case of internal tangency, carries S_1 into S_2 and carries A_1 into A_2; the central similarity transformation with center M_2 and coefficient $k_2 = \mp r_3/r_2$ (minus for external tangency and plus for internal tangency) carries S_2 into S_3 and carries A_2 into A_3; $\cdots$; the central similarity transformation with center M_{n-1} and coefficient $k_{n-1} = \mp r_n/r_{n-1}$ carries S_{n-1} into S_n and carries A_{n-1} into A_n. Finally, the central similarity transformation with center M_n and coefficient $k_n = \mp r_1/r_n$ carries S_n into S_1 and carries the point A_n into the point A_{n+1}. Since

$$k_1 k_2 \cdots k_{n-1} k_n = \left(\mp \frac{r_2}{r_1}\right)\left(\mp \frac{r_3}{r_2}\right) \cdots \left(\mp \frac{r_n}{r_{n-1}}\right)\left(\mp \frac{r_1}{r_n}\right) = \pm 1,$$

the sum of all these central similarities is either a parallel translation (possibly through zero distance, that is, the identity transformation), or it is a central similarity with coefficient -1, that is, a half turn (see Volume One, Chapter One, Section 2, page 21). But we also know that the sum of these n central similarity transformations carries the circle S_1 into itself, and therefore the sum of these transformations must be either the identity transformation or a half turn about the center of S_1. The first case will obtain when $k_1 k_2 \cdots k_n = +1$ and this will happen when the number of external tangencies is even, in particular if all the tangencies are external and there is an even number of circles altogether. The second case will occur when the number of external tangencies is odd, in particular when all the tangencies are external and there is an odd number of circles altogether. This establishes the results stated in parts (a), (b) and (d).

Clearly, if we perform all our transformations in the same order a second time, we obtain the identity transformation. Therefore if A_{2n+1} is the point of S_1 obtained from A_{n+1} in the same way that A_{n+1} was obtained from A_1, then A_{2n+1} always coincides with A_1. This establishes the result stated in part (c).

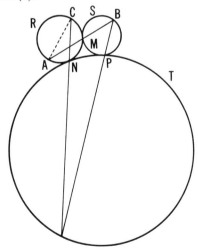

Figure 90

24. First consider three circles R, S, and T, where R and S are externally tangent at M, S and T are externally tangent at P, and T and R are externally tangent at N (Figure 90). Assume also that the radius of T is very much larger than the radii of R and S. If the radius of T grows beyond all bounds, then Figure 90 will approximate Figure 22 more and more closely. In the limit Figure 90 becomes Figure 22, and the result of Problem 23 (a) becomes the desired result for the present problem.

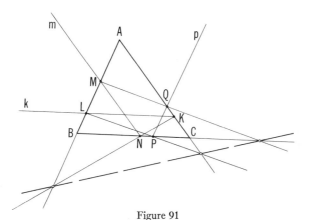

Figure 91

25. Clearly the three triangles ALK, MBN, QPC, the corresponding sides of which are parallel, can be obtained one from another by central similarity transformations. The similarity center of triangles ALK and MBN is the point of intersection of AB and KN, the similarity center of MBN and QPC is the point of intersection of BC and MQ, the similarity center of QPC and ALK is the point of intersection of CA and PL (Figure 91). The desired result now follows from this and from the theorem on the three centers of similarity (see page 29).

The precise formulation of the theorem on the three centers of similarity allows us to improve the result stated in Problem 25: *either the points of intersection of AB and KN, of BC and MQ, and of CA and PL all lie on one line; or exactly one of these points of intersection does not exist, and the corresponding pair of parallel lines is also parallel to the line joining the two remaining points of intersection,* for example, $AB \parallel KN \parallel UV$, where U is the point of intersection of BC and MQ and V is the point of intersection of CA and PL; *or none of the points of intersection exists, that is,* $AB \parallel KN$, $BC \parallel MQ$, and $CA \parallel PL$.

26. (a) Triangle EFD is obtained from the given triangle ABC by a central similarity with center at the centroid of $\triangle ABC$ and with co-efficient $k_1 = -1/2$ [compare the solution to Problem 14 (a)–(c)]; $\triangle LMK$ is obtained from $\triangle EFD$ by a central similarity with center P and coefficient $k_2 = 2$ (Figure 92). Since $k_1 k_2 = -\frac{1}{2} \cdot 2 = -1$, $\triangle LMK$

is obtained from $\triangle ABC$ by a central similarity with coefficient $k = -1$ that is, by a half turn about some point Q; this completes the proof.

(b) From the formula derived on page 29 we have

$$O_1O = \frac{k_2 - 1}{k_1 k_2 - 1} O_1 O_2 .$$

If we let

$$O_1 = N, \quad O_2 = P, \quad O = Q, \quad \text{and} \quad k_1 = -\tfrac{1}{2}, \quad k_2 = 2,$$

then we obtain

$$NQ = \frac{2 - 1}{-1 - 1} NP = -\tfrac{1}{2}NP,$$

that is, Q is obtained from P by a central similarity with center N and coefficient $-\tfrac{1}{2}$. Therefore when P describes a circle S, Q will also describe a circle obtained from S by the indicated central similarity.

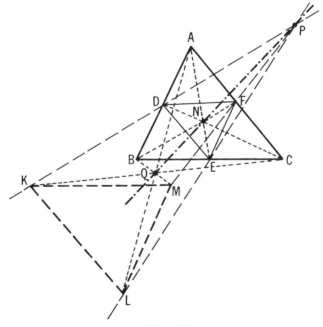

Figure 92

27. (a) Let F_A be some figure containing the vertex A of the triangle; let F_C be obtained from F_A by a central similarity with center P and coefficient $k_1 = PC/PA$; let F_B be obtained from F_C by a central similarity with center N and coefficient $k_2 = NB/NC$ (Figure 93a).

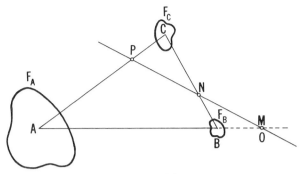

Figure 93a

Clearly the point A of figure F_A corresponds to the point C of figure F_C which, in turn, corresponds to point B of figure F_B. By the theorem on the addition of central similarities, F_B is obtained from F_A by a central similarity with center O and coefficient k; also, O lies on the line BA (since B and A are corresponding points of figures F_B and F_A) and on the line PN (by the theorem on the three centers of similarity), and

$$k = k_1 k_2 = \frac{PC}{PA} \cdot \frac{NB}{NC}.$$

If $k_1 k_2 = 1$, then F_B is obtained from F_A by a translation.
If

$$\frac{AM}{BM} \cdot \frac{BN}{CN} \cdot \frac{CP}{AP} = 1, \quad \text{then} \quad \frac{MB}{MA} = \frac{PC}{PA} \cdot \frac{NB}{NC} = k_1 k_2 = k;$$

since $OB/OA = k$, it follows that M coincides with O and, therefore, lies on line PN. Conversely, if the points M, N and P are collinear, then M is the point of intersection of lines AB and PN and, therefore, coincides with O; therefore

$$\frac{MB}{MA} = \frac{OB}{OA} = k = k_1 k_2 = \frac{PC}{PA} \cdot \frac{NB}{NC}$$

and so

$$\frac{AM}{BM} \cdot \frac{BN}{CN} \cdot \frac{CP}{AP} = 1.$$

Remark 1. Note that in the solution of Problem 27 (a) it was enough to prove just the necessity or just the sufficiency of the conditions of the problem; the other half then follows from what has been proved. Indeed, suppose, for example, that we have proved that if $(AM/BM)(BN/CN)(CP/AP) = 1$, then the points M, N and P lie on a line. To obtain the converse we suppose that M, N and P are collinear, and then must show that $(AM/BM)(BN/CN)(CP/AP) = 1$. To do this let $\bar{M}$ be a point on the line AB such that $(AM/BM)(BN/CN)(CP/AP) = 1$; then by the theorem that we are supposing already proved, the points $\bar{M}$, N and P are collinear; it follows that $\bar{M}$ coincides with M and therefore $(AM/BM)(BN/CN)(CP/AP) = 1$. In a similar manner one can deduce the sufficiency of the condition from its necessity; we leave this deduction to the reader.

Remark 2. The theorem of Menelaus can be connected with the following generalization of Problem 15 of Section 2, Chapter 1, Volume One (page 28): *construct an n-gon $A_1A_2 \cdots A_n$, given the n points M_1, M_2, $\cdots$, M_n that divide its sides in known ratios (positive or negative) $A_1M_1/A_2M_1 = k_1$, $A_2M_2/A_3M_2 = k_2$, $\cdots$, $A_nM_n/A_1M_n = k_n$* (here there is no need to suppose that the number of sides of the n-gon is odd). Just as in the second solution to Problem 15 it can be shown that if $k_1k_2 \cdots k_n \neq 1$, then the problem always has a unique solution; if $k_1k_2 \cdots k_n = 1$, then, in general, the problem has no solution, and for certain special positions of the points M_1, M_2, $\cdots$, M_n the solution to the problem will be undetermined. It follows that if $k_1k_2 \cdots k_n \neq 1$, then the distribution of the points M_1, M_2, $\cdots$, M_n in the plane is arbitrary; on the other hand, if

$$k_1k_2 \cdots k_n = \frac{A_1M_1}{A_2M_1} \frac{A_2M_2}{A_3M_2} \cdots \frac{A_nM_n}{A_1M_n} = 1,$$

then this distribution must satisfy some special conditions. For $n = 3$, by the theorem on the three centers of similarity, these special conditions come down to the requirement that the points M_1, M_2 and M_3 must lie on a line; the theorem of Menelaus then follows.

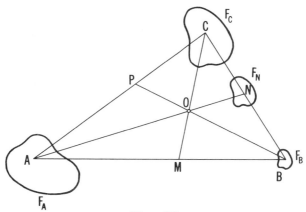

Figure 93b

(b) Let the lines CM, AN and BP meet in a common point O (Figure 93b). Consider any figure F_A containing the point A; let the central similarity with similarity center O and coefficient $k = ON/OA$ carry it into a figure F_N (the point A of figure F_A corresponds to the point N in the figure F_N); let the central similarity with center B and coefficient $k_1 = BC/BN$ carry F_N into F_C (the point N of figure F_N corresponds to the point C in figure F_C); the central similarity with center C and coefficient $k_2 = CB/CN$ carries F_N into F_B (the point N of figure F_N

corresponds to the point B of figure F_B). By the theorem on the addition of central similarities, figure F_C is centrally similar to F_A ; also the center of similarity lies on line CA (C and A are corresponding points of figures F_C and F_A) and on line BO (by the theorem on three centers of similarity), that is, it coincides with P, and the coefficient of similarity is equal to $kk_1 = (ON/OA)(BC/BN)$. Similarly it is shown that F_B and F_A are centrally similar with center M and coefficient $kk_2 = (ON/OA)(CB/BN)$. Thus we have

$$\frac{PC}{PA} = \frac{ON}{OA} \cdot \frac{BC}{BN}, \qquad \frac{MB}{MA} = \frac{ON}{OA} \cdot \frac{CB}{CN};$$

dividing the first equation by the second we have:

$$\frac{PC}{PA} \cdot \frac{MA}{MB} = -\frac{CN}{BN} \quad \text{or} \quad \frac{MA}{MB} \cdot \frac{NB}{NC} \cdot \frac{PC}{PA} = -1,$$

which was to be proved.

Now suppose that $(MA/MB)(NB/NC)(PC/PA) = -1$ and, for example, that lines AN and BP meet in the point O. If $\bar{M}$ is the point of intersection of CO with AB then, by what was proved above, $(\bar{M}A/\bar{M}B)(NB/NC)(PC/PA) = -1$, that is, $\bar{M}$ coincides with M. Thus we see that if $(MA/MB)(NB/NC)(PC/PA) = -1$, then either AN, BP and CM are concurrent or parallel.

Assume, finally, that AN, BP and CM are all parallel; let $\bar{M}$ be a point on line AB such that $(\bar{M}A/\bar{M}B)(NB/NC)(PC/PA) = -1$. In this case $C\bar{M}$ cannot meet AN or BP (for otherwise all three lines AN, BP and $C\bar{M}$ would be concurrent); therefore $\bar{M}$ coincides with M and $(MA/MB)(NB/NC)(PC/PA) = -1$.

Remark. It is not hard to see that the different proofs of Ceva's theorem amount to two applications of the theorem of Menelaus, first to triangle ANC (with the points on the sides being P, O and B) and then to ANB (with the points on the sides being M, O and C).

28. (a) By the theorem of Problem 18 (a) the quadrilaterals $A_1A_2A_3A_4$ and $M_1M_2M_3M_4$, where M_1, M_2, M_3 and M_4 are the centroids of triangles $A_2A_3A_4$, $A_1A_3A_4$, $A_1A_2A_4$ and $A_1A_2A_3$, are centrally similar with similarity coefficient $-\frac{1}{3}$ (and with similarity center at the centroid N of quadrilateral $A_1A_2A_3A_4$). Further, by the result of Problem 14 (a) the quadrilaterals $M_1M_2M_3M_4$ and $H_1H_2H_3H_4$, where H_1, H_2, H_3, H_4 are the orthocenters of the same triangles, are centrally similar with similarity center at the center O of circle S and with similarity coefficient $3/1$. Thus quadrilateral $H_1H_2H_3H_4$ can be obtained from $A_1A_2A_3A_4$ by means of two consecutive central similarities with coefficients $k_1 = -\frac{1}{3}$ and $k_2 = 3$; but the sum of these two central similarities is a central similarity with coefficient $k_1k_2 = -1$, that is, a reflection in a point, and so we obtain the assertion of Problem 33 (a) of Volume One.

Remark. It follows from the theorem on three centers of similarity that the point H of Problem 34 (a) of Volume One lies on a common line with the centroid N and the center O of the circumscribed circle; it is not difficult to show that N is the midpoint of OH (see, for example, page 29).

(b) The solution to part (b) is similar to that of part (a). It is necessary to use the fact that the center $\bar{O}$ of the nine point circle of the triangle lies on a common line with the center O of its circumscribed circle and with the orthocenter H, and that $O\bar{O}/OH = \frac{1}{2}$ (see the solution to Problem 17 (a)).

29. (a) Consider the four points A_1, A_2, A_3 and H_4. From these one can form the four triangles $A_1A_2A_3$, $A_1A_2H_4$, $A_1A_3H_4$ and $A_2A_3H_4$. We shall show that the Euler circles of these triangles all coincide. Indeed the radii of the Euler circles of triangles $A_1A_2A_3$ and $A_2A_3H_4$ are equal to one half the radii of the circles circumscribed about these triangles; thus they are equal, since the circles circumscribed about triangles $A_1A_2A_3$ and $A_2A_3H_4$ are symmetric with respect to the line A_2A_3 [see the solution to Problem 33 (b), Volume One]. Further, the center of the first Euler circle lies at the midpoint of the segment H_4O, where O is the center of S; the center of the second circle is at the midpoint of segment A_1O_1, where O_1 is the center of the circle circumscribed about triangle $A_2A_3H_4$ (because A_1 is the orthocenter of $\triangle A_2A_3H_4$). And since the midpoints of these segments coincide [the quadrilateral $A_1H_4O_1O$ is a parallelogram; see the solution to Problem 33 (c), Volume One], it follows that the centers of the Euler circles coincide, and therefore the circles themselves coincide. In the same way it can be shown that the Euler circles of triangles $A_1A_2H_4$, $A_1A_3H_4$ and $A_1A_2A_3$ coincide.

In a similar manner it can be shown that each of the 32 Euler circles under consideration coincides with the Euler circle of one of the triangles $A_1A_2A_3$, $A_1A_2A_4$, $A_1A_3A_4$, $A_2A_3A_4$, $H_1H_2H_3$, $H_1H_2H_4$, $H_1H_3H_4$, and $H_2H_3H_4$.

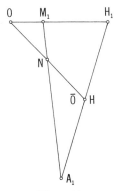

Figure 94

(b) The congruence of all these Euler circles follows from the results of Problems 18 (b) of this volume and 34 (a) of Volume One. By Problem 18 (b), four of these circles meet in a common point $\bar{O}$ and four meet in $\bar{O}'$. Further, the points $\bar{O}$ and $\bar{O}'$ are symmetric with respect to the midpoint H of segment A_1H_1 [see Problem 34 (a), Volume One]; $\bar{O}$ lies on the extension of ON past the point N, while $ON = N\bar{O}$ [N is the point that divides the segment A_1M_1, where M_1 is the centroid of triangle $A_1A_2A_3$, in the ratio $A_1N:NM_1 = 3:1$; see Problem 18 (c)]. From this and the fact that $OM_1 : M_1H_1 = 1:2$ [see Problem 14 (a)], it follows that $\bar{O}$ coincides with H (see Figure 94), and therefore $\bar{O}'$ coincides with $\bar{O}$.

(c) This follows easily from the results of Problems 34 (a) of Volume One and 28 (b) of the present volume.

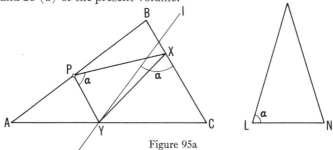

Figure 95a

30. (a) Assume that triangle PXY has already been constructed (Figure 95a). The point Y is obtained from X by a spiral similarity with rotation center P, rotation angle α equal to the angle L of triangle LMN, and similarity coefficient k equal to the ratio of sides LN/LM of this triangle. It follows from this that Y lies on the line l, obtained from BC by a spiral similarity with center P, angle α and coefficient k, and since it lies on side AC, Y is the point of intersection of l and CA. If l is parallel to CA then the problem has no solution; if l coincides with CA then the solution is undetermined.

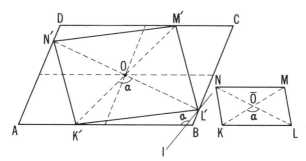

Figure 95b

(b) Note that if parallelogram $K'L'M'N'$ is inscribed in parallelogram $ABCD$ (Figure 95b), then the points of intersection of the diagonals (centers) O' and O of the two parallelograms coincide: for, the common midpoint of diagonals $K'M'$ and $L'N'$ lies on both middle lines[T] of parallelogram $ABCD$, that is, it coincides with the center O.

Now suppose that $K'L'M'N'$ is the desired parallelogram; in this case triangle $K'O'L'$ is similar to triangle $K\bar{O}L$, where $\bar{O}$ is the center of $KLMN$. The spiral similarity with center O, rotation angle $K\bar{O}L$ and similarity coefficient $\bar{O}L/\bar{O}K$ carries side AB of parallelogram $ABCD$ into a line l, whose intersection with the line BC determines vertex L' of the desired parallelogram [compare the solution to part (a)].

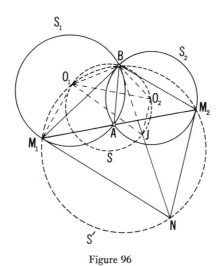

Figure 96

31. Join the second point B of intersection of S_1 and S_2 to the points M_1 and M_2, O_1 and O_2 (Figure 96). Triangle BM_1M_2 is similar to triangle BO_1O_2 (because $\angle BO_1O_2 = 1/2\angle BO_1A = \angle BM_1A$, and

$$\angle BO_2O_1 = 1/2\angle BO_2A = \angle BM_2A);$$

consequently, $\triangle BM_1M_2$ is obtained from $\triangle BO_1O_2$ by a spiral similarity with center B, rotation angle α equal to $\angle M_1BO_1$, and similarity coefficient $k = BM_1/BO_1$. Circumscribe circles S and S' about triangles BO_1O_2 and BM_1M_2; since

---

[T] A middle line of a parallelogram is a line joining the midpoints of two opposite sides.

$\measuredangle BM_1N + \measuredangle BM_2N$

$= (\measuredangle BM_1M_2 + \measuredangle BM_2M_1) + (\measuredangle NM_1M_2 + \measuredangle NM_2M_1)$

$= (\measuredangle BM_1M_2 + \measuredangle BM_2M_1) + (\measuredangle M_1BA + \measuredangle M_2BA)$

$= \measuredangle BM_1M_2 + \measuredangle BM_2M_1 + \measuredangle M_1BM_2 = 180°,$

we see that S' passes through the point N; since

$$\measuredangle O_1BO_2 + \measuredangle O_1JO_2 = \measuredangle M_1BM_2 + \measuredangle M_1NM_2 = 180°,$$

we see that S passes through the point J. Further, we have

$$\measuredangle NBM_1 = \measuredangle NM_2M_1, \qquad \measuredangle JBO_1 = \measuredangle JO_2O_1 ;$$

thus the difference of angles NBM_1 and JBO_1 is equal to the difference of angles NM_2M_1 and JO_2O_1. This difference is the angle between the segments M_2M_1 and O_2O_1 related by the previously indicated spiral similarity, with the same direction $M_2N \parallel O_2J$. Therefore the difference of the angles NBM_1 and JBO_1 is equal to the angle between M_2M_1 and O_2O_1, that is, the angle of rotation α. But from the fact that

$$\measuredangle NBM_1 - \measuredangle JBO_1 = \alpha = \measuredangle M_1BO_1 ,$$

it follows that line NJ passes through B, which proves the first assertion of the problem.

To prove the second assertion it is sufficient to observe that $JO_1 \parallel NM_1 \perp O_1M_1$ and $\measuredangle JNM_1 = \measuredangle M_1M_2B = \measuredangle O_1O_2B$; thus we see that the segment JN forms with the line O_1M_1 an angle of $90° - \measuredangle O_1O_2B$, and the orthogonal projection of JN onto this line is the segment O_1M_1 of constant length r_1 (r_1 is the radius of S_1). Therefore

$$JN = r_1 \cos (90° - \measuredangle O_1O_2B) = r_1 \sin \measuredangle O_1O_2B ,$$

and, clearly, this does not depend on the choice of the line M_1AM_2.

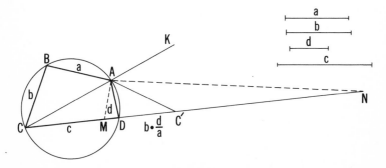

Figure 97a

32. (a) Suppose that the quadrilateral $ABCD$ has been constructed (Figure 97a). The spiral similarity with center A, similarity coefficient d/a and angle of rotation BAD carries $\triangle ABC$ into $\triangle ADC'$, where C' lies on the extension of CD (because $\sphericalangle ADC + \sphericalangle ABC = 180°$). In triangle ACC' we know the segments $CD = c$, $DC' = bd/a$, $DA = d$ and the ratio of the sides $AC'/AC = d/a$; therefore the triangle can be constructed. [Lay off segment CC'. The bisectors of the interior and exterior angles of triangle ACC' at the vertex A meet the line CC' in points M and N such that $C'M/CM = C'N/CN = d/a$; these points can be constructed. Since $\sphericalangle MAN = 90°$, it follows that A is the point of intersection of the circle having the segment MN as diameter with the circle having center D and radius d.] Constructing on segment AC the triangle ABC with sides $AB = a$ and $CB = b$, we have the desired quadrilateral.

The problem has either exactly one solution or has no solution at all.

(b) The solution is similar to that of part (a). Suppose that the quadrilateral $ABCD$ has been constructed (Figure 97b); the spiral similarity with center A, similarity coefficient d/a and rotation angle BAD carries triangle ABC into triangle ADC', where the point C' is determined by the fact that $\sphericalangle CDC' = \sphericalangle B + \sphericalangle D$ and $DC' = b \cdot d/a$. Constructing triangle CDC', we can find A as the intersection of the circle having segment MN as diameter (where M and N are two points on the line CC' such that $C'M/CM = C'N/CN = d/a$), and the circle with center at D and radius d.

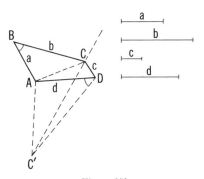

Figure 97b

33. Denote the second point of intersection of circles R and S by N, and join N to A and to B (Figure 98). Note that $\sphericalangle BAN$ and $\sphericalangle ABN$ do not depend on the choice of the line l: Indeed, $\sphericalangle BAN = \sphericalangle MAN$ and is determined by the fixed chord MN; similarly for $\sphericalangle ABN = \sphericalangle MBN$. Hence the third angle, $\sphericalangle ANB = \phi$, of triangle ABN does not depend on the choice of l. (This reasoning is not complete. In our figure we have considered only the case where M lies between A and B. However, if A were on the other side of MN, then A would lie between B and M; if B were on the other side of MN, then B would lie between A and M. We leave it to the reader to complete the reasoning by considering these two cases.) Note that the angle ϕ can be found in terms of the angle α between the two circles R and S (see the statement of Problem 34, page 39 for the definition of the angle between two circles). Indeed, since the angles do not depend on the choice of the line l, choose $l \perp MN$. Then NA and NB coincide with the diameters NO_1A and NO_2B where O_1, O_2 are the centers of R and S, and so either $\phi = \sphericalangle O_1NO_2 = \alpha$, or $\phi = 180° - \alpha$.

Further, since the angles of $\triangle ANB$ do not depend on the choice of l, the same is true for the ratio $NB/NA = k$. Thus, B is obtained from A by a spiral similarity with center N, rotation angle ϕ, and similarity coefficient k. (It is easy to see that the similarity coefficient $k = NB/NA$ is equal to the ratio r_2/r_1 of the radii of R and S: this follows by choosing $l \perp MN$, in which case $NB = 2r_2$ and $NA = 2r_1$. It also follows from the fact that the spiral similarity with coefficient k carries the circle R of radius r_1 onto the circle S of radius r_2.)

It is now clear that if the point Q divides the segment AB in the fixed ratio $AQ/QB = m/n$, then the shape of triangle ANQ does not depend on the choice of the line l; in other words neither the angle $\sphericalangle ANQ = \phi_1$ nor the ratio $NQ/NA = k_1$ depends on l. Thus Q is obtained from A by a spiral similarity with center N, rotation angle ϕ_1, and similarity coefficient k_1. Therefore the locus of all such points Q is a $circle$, namely the circle obtained from the circle R—the locus of the point A—by the spiral similarity just specified.

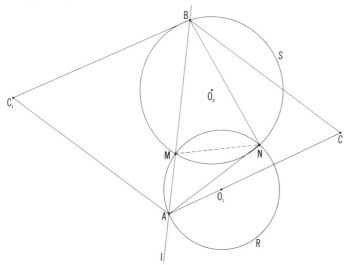

Figure 98

In almost exactly the same way we can solve the problem of the locus of the vertex C of the equilateral triangle ABC, constructed on the segment AB as base, and lying always on the same side of l (either it always lies on the same side of l as N, or it always lies on the opposite side from N). Here also the shape of $\triangle ANC$ does not depend on the choice of l. Therefore the required locus is a *circle*, obtained from the circle R by a certain spiral similarity. If, as in the statement of the problem, we do not require that the point C always lie on the same side of l, then the required locus consists of *two circles*: one of them is the locus of points C that lie on the same side of l as N, and the other is the locus of points C_1 that lie on the opposite side of l (see Figure 98). (If the circles R and S are congruent and if $\sphericalangle O_1NO_2 = 60°$, then the circle described by C_1 reduces to the *point N*.)

The solution to part (c) is similar. Construct the segment NP_1 parallel to, and of the same length and the same direction as AB. Clearly the angle $\sphericalangle ANP_1 = \phi_2$ and the ratio $NP_1/NA = k_2$ do not depend on the choice of l. Therefore the locus of such points P_1 is the *circle* obtained from R by the spiral similarity with center N, rotation angle ϕ_2, and similarity coefficient k_2. This circle is congruent to the circle that would be obtained if we laid off the segment OP parallel, equal and in the same direction as AB from an arbitrary fixed point O and considered the locus of points P obtained in this manner.

Remark. The crux of the solution to Problem 33 (a)–(c) was the fact that *the circles R and S were obtained from each other by a spiral similarity* that carried the point A of R into the point B of S. In the same way we can show that *if R is any figure and if S is obtained from R by a spiral similarity that carries the point A of R into the point B of S then the locus of the*
a) *points Q that divide the segment AB in a given ratio* $AQ:QB = m:n$;
b) *vertices C of equilateral triangles ABC, constructed on the segment AB,* and such that angle $\sphericalangle BAC$ has not only the constant value 60°, but has also a constant direction of rotation (from the ray AB to the ray AC);
c) *endpoints P of segments OP laid off from a fixed point O, and parallel to, equal to, and having the same direction as the segment AB is a figure similar to R and S.†*
The proof of this proposition is just the same as the solution to Problem 33. This remark will be useful to us later.

34. Assume that the problem has been solved (Figure 99). The spiral similarity with center at the point N of intersection of $\bar{S}$ with S, rotation angle α, and coefficient of similarity equal to the ratio of the radii of circles $\bar{S}$ and S, carries S into $\bar{S}$. Under this transformation l_1 and l_2 are carried into lines l_1' and l_2' tangent to $\bar{S}$ and such that l_1' and l_1 form an angle α, and l_2' and l_2 form an angle α; the point M of intersection of l_1

† Or reduces to a point.

and l_2 is carried into the point M' of intersection of l_1' and l_2'.† Consequently, $\sphericalangle MNM' = \alpha$.

We are led to the following construction: Draw the tangents l_1' and l_2' to $\bar{S}$ that form an angle α with l_1 and l_2 respectively; let M' be the point of intersection of l_1' and l_2', and let N be the point of intersection of $\bar{S}$ with the other circle that has MM' as a chord and subtends the angle α on this chord. The spiral similarity with center N, rotation angle α, and similarity coefficient NM'/NM carries $\bar{S}$ into the desired circle S. The problem can have up to eight solutions.

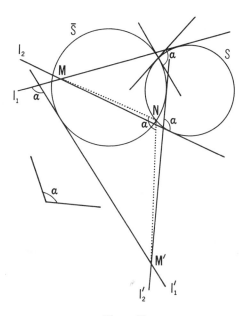

Figure 99

35. Let A, B, C, D, E, F be the points of intersection of our lines (see Figure 32 in the text). The center O of the spiral similarity that carries AB into EF can be found as the point of intersection of the circles circumscribed about triangles AEC and BFC, or as the point of intersection of the circles circumscribed about triangles ABD and EFD (cf. Figure 30b and 31, pp. 43, 44). It follows that these four circles intersect in a common point.

† If $l_1 \parallel l_2$, then the solution to the problem is much simpler, for then we can find the radius r of the desired circle immediately; the center of the circle of radius r that cuts $\bar{S}$ at an angle α lies on one of two well-defined circles concentric with $\bar{S}$. In this case the problem can have up to four solutions.

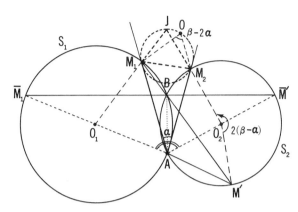

Figure 100

36. (a) We shall show how to obtain the point M_2 from the point M_1. Connect M_1 to the second point B of intersection of S_1 and S_2; let M_1B intersect S_2 in a point M' (Figure 100). Regardless of the position of the given angle α, M' is always obtained from M by the same spiral similarity; this follows from the fact that the shape of triangle M_1AM' does not depend on the position of M_1 (angle AM_1M' is half of arc AB of circle S_1, angle $AM'M_1$ is half of arc AB of circle S_2). By drawing $\bar{M}_1\bar{M}' \perp AB$ (with $A\bar{M}_1$ and $A\bar{M}'$ being diameters of the circles), it is easy to see that the rotation angle β of this spiral similarity is equal to $\sphericalangle O_1AO_2$, and the similarity coefficient is AO_2/AO_1. Moreover, $M'AM_2 = \beta - \alpha,$† $\sphericalangle M'O_2M_2 = 2(\beta - \alpha)$. Thus M_2 is obtained from M_1 by a spiral similarity with center A, rotation angle β and similarity coefficient k (carrying M_1 into M') followed by a rotation with center O_2 and rotation angle $2(\beta - \alpha)$ (carrying M' into M_2). But the sum of these two transformations is a spiral similarity with some center O, with similarity coefficient k, and with rotation angle $-\beta + 2(\beta - \alpha) = \beta - 2\alpha$ (in Figure 100 the directions of rotation from AM_1 to AM' and from O_2M' to O_2M_2 are opposite). Now it only remains to observe that

$$\sphericalangle M_1JM_2 = \sphericalangle AM_1M_2 + \sphericalangle AM_2M_1 = \sphericalangle O_1M_1A + \sphericalangle O_2M_2A - 180°$$

$$= (180° - \sphericalangle M_1AM_2) + (\sphericalangle O_1AM_1 + \sphericalangle O_2AM_2) - 180°$$

$$= (180° - \alpha) + (\beta - \alpha) - 180° = \beta - 2\alpha,$$

and consequently, the circle circumscribed about $\triangle M_1M_2J$ passes through O.

† Or $\alpha - \beta$ (see the second footnote on page 50, Volume One).

(b) The spiral similarity with center A, rotation angle $\beta = \sphericalangle O_1AO_2$, and similarity coefficient $k = AO_2/AO_1$ carries O_1 into O_2; an additional rotation about O_2 through an angle $2(\beta - \alpha)$ leaves O_2 fixed. Therefore

$$\sphericalangle O_1OO_2 = \beta - 2\alpha \qquad \text{and} \qquad \frac{OO_2}{OO_1} = \frac{AO_2}{AO_1}$$

[see the solution to part (a)]. Since the ratio $OO_2/OO_1 = AO_2/AO_1$ is constant, the locus of points O is a circle (see the footnote on page 40); this circle passes through the points A (O coincides with A in case $\alpha = \beta$) and B (O coincides with B in case $\alpha = 0$).

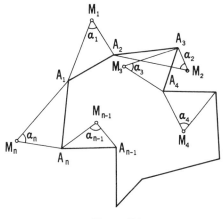

Figure 101

37. Let $A_1A_2 \cdots A_n$ be the desired n-gon (Figure 101). By the hypotheses of the problem we are given the points M_1, M_2, $\cdots$, M_n, we know the angles $A_1M_1A_2 = \alpha_1$, $A_2M_2A_3 = \alpha_2$, $\cdots$, $A_nM_nA_1 = \alpha_n$ and we know the ratios $M_1A_2/M_1A_1 = k_1$, $M_2A_3/M_2A_2 = k_2$, $\cdots$, $M_nA_1/M_nA_n = k_n$. Now perform in succession a spiral similarity with center M_1, rotation angle α_1 and similarity coefficient k_1, a spiral similarity with center M_2, rotation angle α_2 and similarity coefficient k_2 and so forth, ending with a spiral similarity with center M_n, rotation angle α_n and coefficient k_n. Under these transformations the point A_1 will successively occupy the positions A_2, A_3, $\cdots$, A_n and, finally, A_1. Thus A_1 is a fixed point for the sum of the n spiral similarities with centers M_1, M_2, $\cdots$, M_n, with rotation angles α_1, α_2, $\cdots$, α_n and with similarity coefficients k_1, k_2, $\cdots$, k_n.

This sum of spiral similarities represents, in general, a new spiral similarity (with rotation angle $\alpha_1 + \alpha_2 + \cdots + \alpha_n$ and with similarity coefficient $k_1k_2 \cdots k_n$). Since the only fixed point of a spiral similarity is its center, it follows that A_1 must be the center of the resulting spiral similarity. It can be found by performing $n - 1$ times the construction of the center of a spiral similarity which is given as the sum of two known spiral similarities (see pp. 42–43). It is even simpler to find the segment $B'C'$ into which an arbitrary segment BC in the plane is taken by the sum of n spiral similarities, and then to find the center of the spiral similarity carrying BC into $B'C'$ (see pp. 43–44). Having found A_1 there is no difficulty in constructing the remaining n vertices of the n-gon.

If $\alpha_1 + \alpha_2 + \cdots + \alpha_n$ is a multiple of $360°$ and if $k_1k_2 \cdots k_n = 1$, then the sum of the spiral similarities is a *translation*. Since a translation has *no fixed points* at all, the problem has no solution in this case.

It can happen that the sum of the spiral similarities is the *identity transformation* (a translation through zero distance). In this case the problem is undetermined: for the vertex A_1 of the desired n-gon one can choose an *arbitrary point* of the plane.

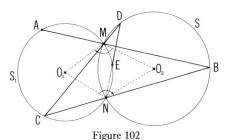

Figure 102

38. (a) The point B is obtained from the point A by a spiral similarity with center N, similarity coefficient $k_1 = r_2/r_1$, where r_1 and r_2 are the radii of the circles S_1 and S_2, and with rotation angle $\angle O_1NO_2$, where O_1 and O_2 are the centers of circles S_1 and S_2 (see the solution to Problem 33). In a similar way the point C is obtained from B by a spiral similarity with center M, coefficient $k_2 = r_1/r_2$, and rotation angle $\angle O_2MO_1$ (Figure 102). The sum of these two central similarity transformations carries A into C. But the sum of two spiral similarities is again a spiral similarity with coefficient $k = k_1k_2 = (r_2/r_1)(r_1/r_2) = 1$, and with rotation angle $\angle O_1NO_2 + \angle O_2MO_1 = 2\angle O_1NO_2$ ($\angle O_2MO_1 = \angle O_1NO_2$ because M and N are symmetric with respect to the line O_1O_2). This spiral similarity is actually an ordinary rotation since the similarity coefficient $k = 1$; the center of this rotation is the center O_1 of S_1, since the rotation carries an arbitrary point A of S_1 to some point C of S_1, that is, it carries S_1 into itself. Finally, just as C was obtained from A by a rotation about O_1 through an angle of $2\angle O_1NO_2$, so too the point E is obtained from C by a rotation about O_1 through an angle of $2\angle O_1NO_2$. Thus E is obtained from A by a rotation about O_1 through an angle of $4\angle O_1NO_2$. The assertion of part (a) follows from the fact that the angle $4\angle O_1NO_2$ does not depend on the position of A.

(b) E will coincide with A if $4\sphericalangle O_1NO_2 = 360°$, that is, if $\sphericalangle O_1NO_2 = 90°$. In other words, E will coincide with A if the two circles S_1 and S_2 are orthogonal, that is, if the angle between S_1 and S_2 is $90°$ (for the definition of the angle between two circles see the statement of Problem 34, page 39).

39. (a) A_2 is obtained from A_1 by a spiral similarity with center N, coefficient $k_1 = r_2/r_1$, and rotation angle $\sphericalangle O_1NO_2$; A_3 is obtained from A_2 by a spiral similarity with the same center N, with coefficient $k_2 = r_3/r_2$, and with rotation angle $\sphericalangle O_2NO_3$; A_4 is obtained from A_3 by a spiral similarity with center N, coefficient $k_3 = r_1/r_3$, and rotation angle $\sphericalangle O_3NO_1$; here O_1, O_2, and O_3 are the centers of the circles S_1, S_2, and S_3, and r_1, r_2, r_3 are their radii (compare the solution to Problem 33). The sum of these three spiral similarities is a translation, because

$$k_1k_2k_3 = \frac{r_2}{r_1}\frac{r_3}{r_2}\frac{r_1}{r_3} = 1 \quad \text{and} \quad \sphericalangle O_1NO_2 + \sphericalangle O_2NO_3 + \sphericalangle O_3NO_1 = 360°.$$

Further, since this translation carries each point A_1 of the circle S_1 into a point A_4 of *the same circle*, that is, it carries S_1 onto itself, it must be the identity transformation. Thus A_4 is obtained from A_1 by the identity transformation, that is, $A_4 = A_1$.

Clearly the result of this problem can be generalized to an arbitrary number of circles that intersect in a common point. Thus, in Figure 103 we show four circles intersecting in a common point. The point A_5 is obtained from the point A_1 by the sum of *four* spiral similarities; but this sum is the identity transformation and so $A_5 = A_1$.

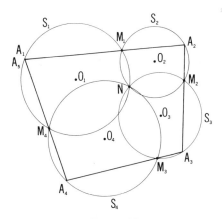

Figure 103

(b) Let O_i and r_i denote the center and radius of the circle S_i, where the index i can take on any of the three values 1, 2, 3. The point A_2 is obtained from A_1 by a spiral similarity with center N_1, coefficient $k_1 = r_2/r_1$, and rotation angle $\angle O_1 N_1 O_2$ (compare the solution to Problem 33). Likewise, the point A_3 is obtained from A_2, A_4 from A_3, A_5 from A_4, A_6 from A_5, and A_7 from A_6 by a spiral similarity with center N_2, coefficient $k_2 = r_3/r_2$, rotation angle $\angle O_2 N O_3$; with center N_3, coefficient $k_3 = r_1/r_3$, rotation angle $\angle O_3 N_3 O_1$; with center M_1, coefficient $k_4 = r_2/r_1 (= k_1)$, and rotation angle $\angle O_1 M_1 O_2$; with center M_2, coefficient $k_5 = r_3/r_2 (= k_2)$, rotation angle $\angle O_2 M_2 O_3$; with center M_3, coefficient $k_6 = r_1/r_3 (= k_3)$, rotation angle $\angle O_3 M_2 O_1$. Thus A_7 is obtained from A_1 by six consecutive spiral similarities.

The sum of the first three of these is an ordinary rotation, since

$$k_1 k_2 k_3 = \frac{r_2}{r_1} \frac{r_3}{r_2} \frac{r_1}{r_3} = 1.$$

This rotation carries an arbitrary point A_1 of the circle S_1 to a point A_4 of the same circle, that is, it carries S_1 into itself. Hence the sum of the first three spiral similarities is a rotation about O_1 through the angle

$$\alpha = \angle O_1 N_1 O_2 + \angle O_2 N_2 O_3 + \angle O_3 N_3 O_1 .$$

In the same way the sum of the last three spiral similarities is also a rotation about O_1, through the angle

$$\beta = \angle O_1 M_1 O_2 + \angle O_2 M_2 O_3 + \angle O_3 M_3 O_1 .$$

The sum of our original six spiral similarities is the same as the sum of these two rotations about O_1, and is therefore again a rotation about O_1, through the angle $\alpha + \beta$. We shall now show that $\alpha + \beta = 0$. Indeed,

$$\angle O_1 N_1 O_2 = - \angle O_1 M_1 O_2$$

$$\angle O_2 N_2 O_3 = - \angle O_2 M_2 O_3$$

$$\angle O_3 N_3 O_1 = - \angle O_3 M_3 O_1 ,$$

and therefore $\alpha = -\beta$ (see Figure 35b).

Thus the sum of our six spiral similarities is a rotation about O_1 through the zero angle, that is, it is the identity transformation. Since this transformation carries A_1 into A_7, we have shown that $A_1 = A_7$.

The result of this problem can be generalized to the case of an arbitrary number of pairwise intersecting circles.

40. (a) First let l be a circle with a radius that is much larger than the radii of circles S_1 and S_2, and apply the result of Problem 39 (b). If we now let the radius of l increase beyond all bounds so that l becomes more and more nearly a straight line intersecting the circles S_1 and S_2, then in the limit we obtain the desired result (compare the solution to Problem 24).

(b) The result of part (a) becomes the result of part (b) if we move the line l so that the points K and L coincide and the points P and Q coincide, that is, if we move l so that it becomes a common tangent to S_1 and S_2.

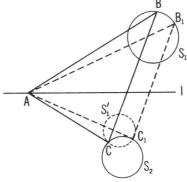

Figure 104

41. Suppose that $\triangle ABC$ has been constructed (Figure 104). The point B is carried into the point C by the dilative reflection with center A, axis l and similarity coefficient n/m. Therefore C lies both on the circle S_2 and on the circle S_1' obtained from S_1 by this dilative reflection (Figure 104). The problem can have two, one, or no solutions.

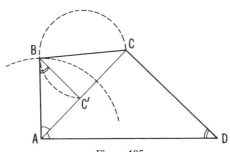

Figure 105

42. Assume that the quadrilateral $ABCD$ has been constructed. The dilative reflection with center A, axis AC and similarity coefficient AB/AD carries $\triangle ADC$ into $\triangle ABC'$ (Figure 105).

(a) AC and $AC' = AC (AB/AD)$ are known; consequently we can lay off these segments on some line. Further, $\sphericalangle ABC' = \sphericalangle ADC$; therefore

$$\sphericalangle C'BC = \sphericalangle ABC - \sphericalangle ADC = \sphericalangle B - \sphericalangle D$$

is known to us; therefore B can be found as the point of intersection of the circular arc constructed on the chord CC' and subtending an angle equal to $\angle B - \angle D^{\mathrm{T}}$ with the circle having center A and radius equal to AB. It is now easy to find vertex D. The problem has at most one solution.

(b) Since sides BC and $BC' = DC(AB/AD)$ and

$$\angle C'BC = \angle B - \angle D$$

of $\triangle CBC'$ are known, this triangle can be constructed. Vertex A can be found as that point on line CC' for which $AC'/AC = AB/AD$. The problem has a unique solution.

(c) In this case we know the ratios

$$\frac{BC}{BC'} = \frac{BC}{CD \cdot (AB/AD)} = \frac{BC}{CD} \cdot \frac{AD}{AB}.$$

The point B can be found as the point of intersection of the locus of points for which the ratio of the distances to C and to C' has the given value $(BC/CD)(AD/AB)$ (see the footnote on page 40), with the circle of center A and radius AB. The problem can have at most one solution.

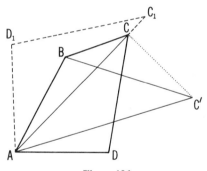

Figure 106

43. The dilative reflection with center A, axis AC and similarity coefficient AB/AD followed by a rotation with center A and rotation angle γ carries triangle ADC into triangle ABC' (see Figure 106). The remainder of the construction proceeds in a manner analogous to the solution of Problem 42. [In the solution of the problem analogous to Problem 42 (b), vertex A is found as the intersection of the locus of points the ratio of whose distances from C' and C has the fixed value AB/AD, and the arc constructed on the chord CC' and subtending the angle α.]

T For the details of the construction of the arc see, for example, *Hungarian Problem Book I* in this series, Problem 1895/2, Note, p. 30.

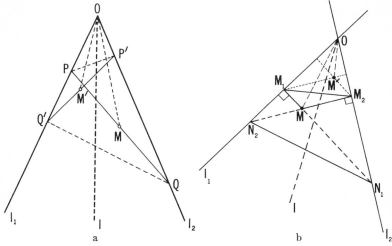

Figure 107

44. (a) Let l_1 and l_2 meet at O and form an angle α there (Figure 107a). [If $l_1 \parallel l_2$, then the problem makes no sense: in this case segment PQ will only exist when the point M is equidistant from l_1 and l_2, and hence M cannot describe a circle.] Triangles OPP' and OQQ' are similar (right triangles with equal base angles); consequently, $OP/OP' = OQ/OQ'$. Therefore $OP/OQ = OP'/OQ'$, that is, triangles OPQ and $OP'Q'$ are also similar; hence triangles OPM and $OP'M'$ are similar (OM and OM' are the medians of OPQ and $OP'Q'$). This means that the ratio $OM'/OM = OP'/OP = \cos\alpha$ does not depend on the choice of the point M, and $\sphericalangle M'OP' = \sphericalangle MOP$, that is, OM and OM' form equal angles with the bisector l of angle POQ. Thus M' is obtained from M by a dilative reflection with center O, axis l and coefficient $k = \cos\alpha$; therefore, when M describes the circle S, M' describes the image circle S'.

(b) First of all, it is clear that if $\sphericalangle M_1OM_2 = \alpha = 90°$, then each point M of the plane is taken into the same point $M' = O$; thus the only interesting case is when $\alpha \neq 90°$. Denote the point of intersection of MM_1 with l_2 by N_1, and that of MM_2 with l_1 by N_2 (Figure 107b). Just as in the solution to part (a), we show that the triangles OM_1M_2 and ON_1N_2 are similar with similarity coefficient $k = OM_1/ON_1 = \cos\alpha$ [if M_1 and M_2 play the roles of the points P and Q of part (a), then the points P' and Q' of that part play the roles of N_1 and N_2, respectively]. Since M and M' are the points of intersection of the altitudes of the similar triangles ON_1N_2 and OM_1M_2, it follows that $OM'/OM = k = \cos\alpha$, and that $\sphericalangle M'OM_1 = \sphericalangle MOM_2$. From this we conclude that the lines OM and OM' are symmetric with respect to the bisector l of angle M_1OM_2.

Thus the point M' of part (b) is obtained from M by means of the *very same* dilative reflection as was the point M' of part (a); if M describes a circle S, then M' describes the image circle S'.

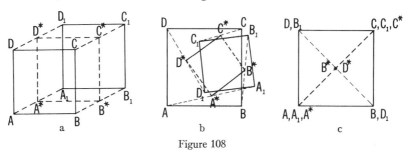

Figure 108

45. (a) If the contours of the two squares $ABCD$ and $A_1B_1C_1D_1$ are traversed in the same direction, that is, if these two squares are *directly similar*, then $A_1B_1C_1D_1$ is obtained from $ABCD$ either by a *spiral similarity* or by a *translation* (see Theorem 1 on page 53). If $A_1B_1C_1D_1$ is obtained from $ABCD$ by a translation and if A^*, B^*, C^*, D^* are the midpoints of the segments AA_1, BB_1, CC_1, DD_1, then $A^*B^*C^*D^*$ is also obtained from $ABCD$ by a translation, in the same direction and through half the distance (Figure 108a).

On the other hand if $A_1B_1C_1D_1$ is obtained from $ABCD$ by a spiral similarity that is not a half-turn (in this case the midpoints A^*, B^*, C^*, and D^* all coincide with the center of rotation), then the midpoints A^*, B^*, C^*, D^* will be the vertices of a square (Figure 108b), because of the result formulated as a Remark following the solution to Problem 33.

However, if the contours of the squares $ABCD$ and $A_1B_1C_1D_1$ are traversed in opposite directions, then the result is no longer valid; for example, we may consider the case when the points $A_1 = A$, $C_1 = C$, $B_1 = D$, and $D_1 = B$. (Figure 108c).

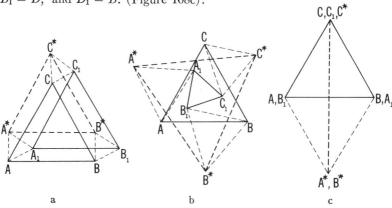

Figure 109

(b) This problem can also be solved on the basis of Theorem 1 and the Remark following the solution to Problem 33 (Figure 109a, b); however, the (simpler!) case when $A_1B_1C_1$ is obtained from ABC by means of a translation requires special consideration. In this case $A^*B^*C^*$ is obtained from ABC by a translation through the same distance but in the direction AA^* (Figure 109a).

If the direction in which the contours ABC and $A_1B_1C_1$ are traversed is the same, and is opposite to the direction in which the three contours AA_1A^*, BB_1B^*, and CC_1C^* are traversed, then the assertion of the problem remains valid. However, in general the assertion is not valid without some hypothesis about the directions in which the contours are traversed; see, for example, Figure 109c.

46. (a) Since the two squares are directly similar figures, it follows from Theorem 1 that $MNPQ$ is obtained from $ABCD$ either by a translation or by a spiral similarity. The assertion of the problem is obvious in the case of a translation, for then the four segments in question, AM, BN, CP, and DQ, all have the same length. Let us therefore assume that $MNPQ$ is obtained from $ABCD$ by a spiral similarity.

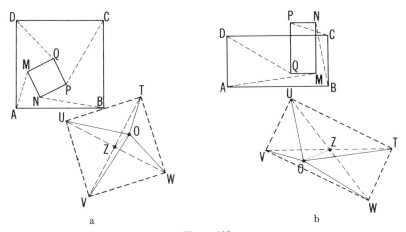

Figure 110

From some point O lay off segments, OT, OU, OV, OW that are equal to, parallel to, and have the same direction as the segments AM, BN, CP, DQ. The four points T, U, V, W will be the vertices of a square (Figure 110a), by the Remark following the solution to Problem 33. Let Z be the center of the square $TUVW$. If we apply the law of cosines to the triangles OTZ and OVZ we have

$$OT^2 = OZ^2 + ZT^2 - 2\,OZ\cdot ZT \cos \sphericalangle OZT$$

and

$$OV^2 = OZ^2 + ZV^2 - 2\,OZ\cdot ZV \cos \sphericalangle OZV$$
$$= OZ^2 + ZT^2 + 2\,OZ\cdot ZT \cos \sphericalangle OZT,$$

from which it follows that

$$OT^2 + OV^2 = 2\,OZ^2 + 2\,ZT^2.$$

In exactly the same way we obtain the formula

$$OU^2 + OW^2 = 2\,OZ^2 + 2\,ZU^2.$$

Therefore, since $ZT = ZU$,

$$OT^2 + OV^2 = OU^2 + OW^2,$$

that is,

$$AM^2 + CP^2 = BN^2 + DQ^2.$$

It is easy to see that this reasoning remains valid when $ABCD$ and $MNPQ$ are any two directly similar rectangles (Figure 110b). However, the result cannot be extended to the case when $ABCD$ and $MNPQ$ are oppositely similar squares or rectangles. Thus, for example, in the notations of Figure 108c, $AA_1 = CC_1 = 0$, but $BB_1 = DD_1 \neq 0$, and so $AA_1^2 + CC_1^2 \neq BB_1^2 + DD_1^2$.

(b) We proceed as in part (a). The second regular hexagon is obtained from the first either by a translation or by a spiral similarity. The assertion of the problem is obvious in the case of a translation, so we assume that the second hexagon is obtained from the first by a spiral similarity.

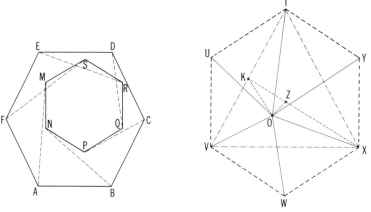

Figure 111

Now from some point O we lay off segments OT, OU, OV, OW, OX, and OY, that are equal to, parallel to, and have the same direction as the segments AM, BN, CP, DQ, ER, and FS. Then by the Remark following the solution to Problem 33, the six points T, U, V, W, X, Y are the vertices of a regular hexagon (Figure 111). Let K denote the midpoint of segment TV, and let Z denote the center of the hexagon

$TUVWXY$, so that Z is also the center of the equilateral triangle TVX. Clearly we have $XZ:ZK = 2:1$. Just as in the solution to part (a), we show that

$$OT^2 + OV^2 = 2\,OK^2 + 2KT^2.$$

Further, applying the law of cosines to triangles OZX and OZK we have:

$$OX^2 = OZ^2 + ZX^2 - 2\,OZ \cdot ZX \cos \sphericalangle OZX$$

and

$$OK^2 = OZ^2 + ZK^2 - 2\,OZ \cdot ZK \cos \sphericalangle OZK.$$

Thus we have

$$OT^2 + OV^2 = 2\,OK^2 + 2KT^2$$

$$= 2\,OZ^2 + 2ZK^2 - 4\,OZ \cdot ZK \cos \sphericalangle OZK + 2KT^2.$$

But $ZK = \frac{1}{2}ZX$ and $\cos \sphericalangle OZK = -\cos \sphericalangle OZX$; therefore

$$4\,OZ \cdot ZK \cos \sphericalangle OZK = -2\,OZ \cdot ZX \cos \sphericalangle OZX,$$

and so

$$OT^2 + OV^2 + OX^2 = 3\,OZ^2 + 2ZK^2 + 2KT^2 + ZX^2 = 3\,OZ^2 + 3ZT^2,$$

where the last equality follows from the fact that

$$2(ZK^2 + KT^2) + ZX^2 = 2ZT^2 + ZX^2 = 3ZT^2.$$

In the same way it can be shown that

$$OU^2 + OW^2 + OY^2 = 3\,OZ^2 + 3ZU^2 = 3\,OZ^2 + 3ZT^2,$$

from which we have

$$OT^2 + OV^2 + OX^2 = OU^2 + OW^2 + OY^2$$

or

$$AM^2 + CP^2 + ER^2 = BN^2 + DQ^2 + FS^2.$$

47. (a) If X and X' are opposite vertices of the desired rectangle (Figure 112), then they lie respectively on the circles S and S' with diameters AB and CD. These circles can be regarded as directly similar figures with corresponding points X and X'. By Theorem 1 (page 53) there exists a spiral similarity (or translation) that carries S into S' and takes the point X into the point X'. This spiral similarity can be determined in the following manner. Pass through the points A and C an arbitrary pair of lines, parallel to one another, and meeting the circles S and S' respectively in points M and M'. Angles MAX and $M'CX'$ are equal, since they are angles with parallel sides; therefore arc MX of circle S has the same angular measure as arc $M'X'$ of circle S'. It follows that our spiral similarity carries the point M into M'; and since it carries the center K of circle S into the center K' of circle S', the problem is reduced to finding the spiral similarity (or translation) carrying a known segment KM into another known segment $K'M'$ (see pages 43–44).

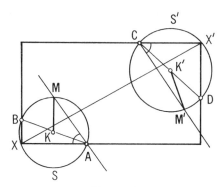

Figure 112

If O is the center, α the rotation angle, and k the similarity coefficient of this spiral similarity, then

$$\triangle OMM' \sim \triangle OXX'$$

because $\sphericalangle MOM' = \sphericalangle XOX' = \alpha$ and $OM'/OM = OX'/OX = k$. Since side XX' of triangle OXX' is known, we can find side OX, and the point X can then be found as a point of intersection of circle S with the circle of center O and radius OX. The problem can have two, one, or no solutions; the points A, B, C, D may lie on the sides of the constructed rectangle, or on their extensions.

A special case occurs when segment KM is taken into segment $K'M'$ by a translation (or, in other words, when the segments AB and CD are equal, parallel, and have the same direction). In this case the problem will have no solution if the magnitude of the translation is not equal to the given length of the diagonal of the rectangle, and otherwise the problem is undetermined (for the vertex X of the desired rectangle one can choose any point of the circle S).

(b) Assume that the quadrilateral has been constructed (Figure 113a). Its vertices B and D lie on circular arcs S and S' constructed on the diagonal AC and subtending angles equal to the given angles B and D. Denote angle BAC by α and angle DCA by β. Then from triangle ABC we have

$$\alpha + \sphericalangle B + \sphericalangle C - \beta = 180°,$$

and so

$$\beta - \alpha = \sphericalangle B + \sphericalangle C - 180°,$$

that is, the quantity $\beta - \alpha$ is known. Suppose for definiteness that $\beta - \alpha > 0$; pass a line through the point C in such a manner that the angle between this line and the diagonal CA is equal to $\beta - \alpha$. Let this line meet the arc S' in a point E; then $\sphericalangle DCE = \beta - (\beta - \alpha) = \alpha$, that is, $\sphericalangle DCE = \sphericalangle BAC$ and, therefore, arc DE = arc BC.

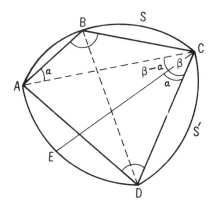

Figure 113a

From this we have the following construction, analogous to the solution to part (a). On the given segment AC construct a circular arc S subtending the given angle B, and on the other side of the segment construct a circular arc subtending the given angle D. Through the point C pass a line forming an angle $\beta - \alpha = \angle B + \angle C - 180°$ with the line AC and meeting the arc S' in a point E. Now it only remains to solve the following problem, to which part (a) was also reduced: on two circles S and S' two points C and E are given; find points B and D on them such that arcs CB and ED have the same angular measure and such that the segment DB has a given length.

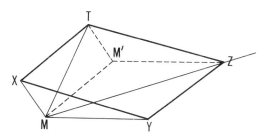

Figure 113b

(c) Let $XYZT$ be the desired parallelogram; MX, MY, MZ, MT are the four given lines (Figure 113b). Translate $\triangle XMY$ in the direction of YZ through a distance equal to the length of YZ so that the segment XY coincides with segment TZ. If M' is the new position of M, then in the quadrilateral $MZM'T$ we know the diagonals ZT and $MM' = YZ$ and the angles $\angle ZMT$, $\angle ZM'T = \angle YMX$, $\angle MTM' = \angle XMT$, $\angle MZM' = \angle YMZ$. Thus the present problem is reduced to part (b).

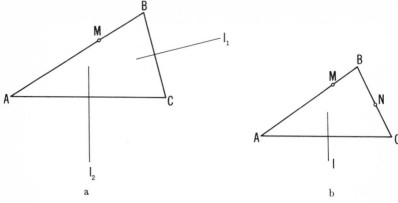

Figure 114

48. (a) Suppose that $\triangle ABC$ has been constructed (Figure 114a). Perform in succession a central similarity with center M and coefficient $-k$ and two reflections, one in the line l_1, the other in l_2; the point A is first taken to B, then B is moved to C and, finally, C is moved to A. Thus A is a fixed point of the sum of the central similarity and two reflections in lines. This sum is a transformation that carries each figure F into a figure F', directly similar to F, and by Theorem 1 it is a *spiral similarity*. It is not difficult to locate the center O of this spiral similarity; for this, it is enough to construct the segment $P'Q'$ that is the image of an arbitrary segment PQ in the plane under the sum of these three transformations, and then find the rotation center of these two segments (see pages 43–44). The vertex A must coincide with the point O (because the only fixed point of a spiral similarity is the center); after this it is easy to find the two other vertices B and C of the desired triangle. If $k = 1$ and $l_1 \perp l_2$, the problem is either impossible or undetermined; in all other cases it has a unique solution (compare the solution to Problem 37).

(b) Suppose that triangle ABC has been constructed (Figure 114b). Carry out in succession central similarities with centers M and N and coefficients $-k_1$ and $-k_2$ and a reflection in the line l. The sum of these transformations carries the point A into itself and thus A is a fixed point of this sum. This sum obviously carries every figure F into a figure F' oppositely similar to F, and so by Theorem 2 it is a *dilative reflection*. It is not difficult to find the axis and the center O of this transformation if we find the segment $P'Q'$ that is the image of an arbitrary segment PQ in the plane under this transformation (see pages 54–55, in particular Figure 40). Then $A = O$. Having constructed A there is no difficulty in finding the remaining vertices B and C of the desired triangle.

If $k_1k_2 = 1$, then our sum of transformations is a glide reflection (or merely a reflection in a line); in this case the problem either has no solution or is undetermined. In all other cases there is a unique solution.

Chapter Two. Further applications of isometries and similarities

49. (a) Many triangles $A_1B_1C_1$ can be constructed, congruent to the given triangle ABC and such that the sides A_1B_1 and A_1C_1 pass through two given points M and N. By Theorem 1 (page 65) the side B_1C_1 of any such triangle must be tangent to some fixed circle S, which can be found by constructing three such triangles, $A_1B_1C_1$, $A_2B_2C_2$ and $A_3B_3C_3$ (Figure 115a). After this it remains to pass a tangent to S from the given point P: side B_0C_0 of the desired triangle $A_0B_0C_0$ will lie on this tangent.

The problem can have two, one, or no solutions.

(b) This problem is very similar to part (a). Many triangles $A_1B_1C_1$ can be constructed congruent to the given triangle ABC and such that A_1B_1 is tangent to the given circle S_1 and A_1C_1 is tangent to the given circle S_2. The third sides of all these triangles will be tangent to some circle S (page 66) which is easy to find by constructing three such triangles, $A_1B_1C_1$, $A_2B_2C_2$ and $A_3B_3C_3$ (Figure 115b). After this it only remains to pass a common tangent to the circle S and to the given circle S_3; side B_0C_0 of the desired triangle $A_0B_0C_0$ will lie on this line.

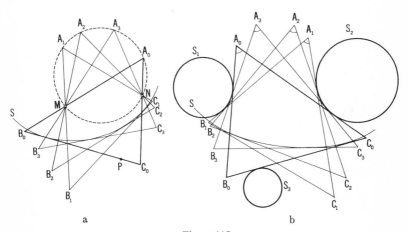

a   b

Figure 115

The circles S and S_3 will have, in general, four common tangents. In addition, triangle $A_1B_1C_1$ (and hence also $A_2B_2C_2$ and $A_3B_3C_3$) can be constructed in the following four essentially different ways:

1°. The circle S_1 and the point C_1 lie on opposite sides of A_1B_1; the circle S_2 and the point B_1 lie on opposite sides of A_1C_1 (Figure 115b).

2°. S_1 and the point C_1 lie on opposite sides of A_1B_1; S_2 and B_1 lie on the same side of A_1C_1.

$3°$. S_1 and C_1 lie on the same side of A_1B_1 ; S_2 and B_1 lie on opposite sides of A_1C_1 .

$4°$. S_1 and C_1 lie on the same side of A_1B_1 ; S_2 and B_1 lie on the same side of A_1C_1 .

Thus the problem can have up to sixteen solutions.

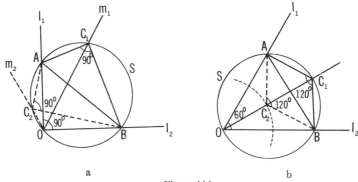

a   b

Figure 116

50. (a) Points C_1 and C_2 in Figure 116a lie on the circle S circumscribed about triangle ABO. By Theorem 2, when the segment AB slides with its end-points on the sides of angle l_1Ol_2, the vertices C_1 and C_2 of triangles ABC_1 and ABC_2 describe lines m_1 and m_2, passing through O. [More accurately, they describe segments on these lines. We leave it to the reader to determine the lengths of these segments for himself.]

(b) The point C_1 in Figure 116b lies on the circle S circumscribed about triangle ABO; the point C_2 is the center of this circle. When the segment AB slides with its endpoints on the sides of angle l_1Ol_2, vertex C_1 of triangle ABC_1 describes a line passing through O, while vertex C_2 of triangle ABC_2 describes a circle with center O (see pp. 66–67). [More precisely, C_1 describes some segment on this line and C_2 describes some arc of the circle.]

51. Assume that triangle ABC has been constructed and consider $\triangle \bar{A}\bar{B}\bar{C}$ with some fixed length of the hypotenuse $\bar{A}\bar{B} = a$, and centrally similar to $\triangle ABC$ with similarity center O at the point of intersection of lines l_1 and l_2. It follows from the solution to Problem 50 (a) that vertex $\bar{C}$ lies on one of the easily constructed lines m_1, m_2, m_3 and m_4 (see Figure 117). (To construct the lines it is enough to construct the right triangles $\bar{A}\bar{B}\bar{C}_1$, $\bar{A}\bar{B}\bar{C}_2$, $\bar{A}\bar{B}\bar{C}_3$ and $\bar{A}\bar{B}\bar{C}_4$ with given acute angle α, such that the vertices at the acute angles are any points whatsoever on the lines l_1 and l_2.) Clearly C lies on this same line. Thus C is found as the point of intersection of one of the lines m_1, m_2, m_3 and m_4 with the circle S.

The problem can have up to eight solutions.

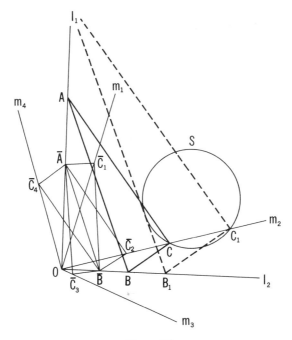

Figure 117

52. (a) Consider the quadrilateral BM_1M_2P, where B is the second point of intersection of S_1 and S_2. When M_1M_2 is rotated about A, triangle BM_1M_2 remains similar to itself (see the solution to Problem 31); therefore the quadrilateral BM_1M_2P also remains similar to itself. Now all the assertions of the problem follow from the facts that the point B of this quadrilateral remains fixed, the vertices M_1 and M_2 describe circles S_1 and S_2, and side M_1M_2 always passes through the fixed point A (see above, page 69). (We assume that the triangle M_1M_2P is constructed on a definite side of M_1M_2—either on the same side as that on which triangle BM_1M_2 lies or on the opposite side; otherwise one must allow P to describe two circles and one must make the corresponding change in the formulation of the problem, concerning the lines M_1P and M_2P.)

(b) Since $\triangle BM_1M_2 \sim \triangle BN_1N_2$ [see the solution to part (a)], it follows that $\triangle BM_1N_1 \sim \triangle BM_2N_2$; consequently, $\sphericalangle BN_1M_1 = \sphericalangle BN_2M_2$ and, therefore, a circle can be circumscribed about quadrilateral BN_1QN_2. But this means that the locus of points Q is the circle Γ (Figure 118) circumscribed about triangle BN_1N_2.

When l_0 is rotated about A, triangle BN_1N_2 changes, but always remains similar to its original position [see the solution to Problem (a)].

Since at the same time the point B remains fixed, and the points N_1 and N_2 describe circles S_1 and S_2, it follows that the center of the circumscribed circle Γ of this triangle describes a circle.

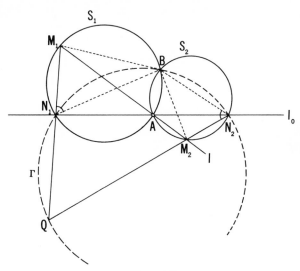

Figure 118

53. Clearly $\sphericalangle ABM = \sphericalangle AO_1O_2$ (they each equal one half of arc AM of circle S_1; see Figure 119; similarly $\sphericalangle ACM = \sphericalangle AO_2O_1$. Therefore when l is rotated around A, $\triangle AO_1O_2$ changes in such a way that it remains similar to itself (and to triangle ABC). Since the point A is fixed, and the points O_1 and O_2 describe lines—the perpendicular bisectors of AB and AC—it follows that the midpoint of the segment O_1O_2 also describes a line.

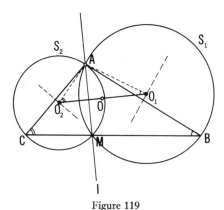

Figure 119

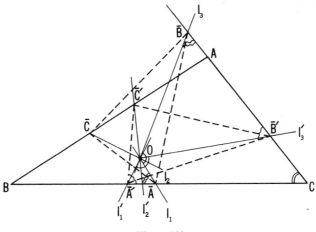

Figure 120

54. Let $\bar{A}\bar{B}\bar{C}$ and $\bar{A}'\bar{B}'\bar{C}'$ be any two positions of $\triangle\bar{A}\bar{B}\bar{C}$ (Figure 120). Since the angle between $O\bar{A}$ and $O\bar{B}$ ($O\bar{A}'$ and $O\bar{B}'$) is equal to the angle between $\bar{A}C$ and $\bar{B}C$, it follows that

$$\sphericalangle\bar{A}O\bar{B} + \sphericalangle\bar{A}C\bar{B} = \sphericalangle\bar{A}'O\bar{B}' + \sphericalangle\bar{A}'C\bar{B}' = 180°,$$

$$\sphericalangle O\bar{A}C + \sphericalangle O\bar{B}C = \sphericalangle O\bar{A}'C + \sphericalangle O\bar{B}'C = 180°,$$

and

$$\sphericalangle O\bar{A}'\bar{A} = \sphericalangle O\bar{B}'\bar{B}, \qquad \sphericalangle O\bar{A}\bar{A}' = \sphericalangle O\bar{B}\bar{B}'.$$

Thus triangles $O\bar{A}\bar{A}'$ and $O\bar{B}\bar{B}'$ are similar; in the same way it can be shown that triangle $O\bar{C}\bar{C}'$ is also similar to them. It follows from this that $\triangle\bar{A}'\bar{B}'\bar{C}'$ is obtained from $\bar{A}\bar{B}\bar{C}$ by a spiral similarity with center O. Thus when the lines l_1, l_2, l_3 are rotated about O, triangle $\bar{A}\bar{B}\bar{C}$ changes but remains similar to itself; each two positions of this triangle have the same rotation center O. From this and from the fact that the vertices of the triangle describe lines, it follows that each of its points describes a line, and this answers the question in part (b). The point O has the same position with respect to all of the triangles $\bar{A}\bar{B}\bar{C}$; therefore to understand its position in these triangles, it is enough to consider just one of them, for example triangle $\bar{A}_0\bar{B}_0\bar{C}_0$ formed from the perpendiculars from O to the sides of ABC.

1°. If O is the center of the circumscribed circle about $\triangle ABC$, then the sides of $\triangle\bar{A}_0\bar{B}_0\bar{C}_0$ are parallel to the sides of $\triangle ABC$ (the mid-lines of the triangle) and O is the point of intersection of the altitudes of $\triangle\bar{A}_0\bar{B}_0\bar{C}_0$.

2°. If O is the center of the inscribed circle of $\triangle ABC$, then $O\bar{A}_0 = O\bar{B}_0 = O\bar{C}_0$, and O is the center of the circumscribed circle about $\triangle\bar{A}_0\bar{B}_0\bar{C}_0$.

$3°$. If O is the point of intersection of the altitudes of $\triangle ABC$, then O is the point of intersection of the bisectors of triangle $\bar{A}_0\bar{B}_0\bar{C}_0$ ($\measuredangle OA_0B_0 = \measuredangle OCB_0$ since these angles cut off the same arc on the circle circumscribed about OA_0CB_0; $\measuredangle OA_0C_0 = \measuredangle OBC_0$ for a similar reason; $\measuredangle OCB_0 = \measuredangle OBC_0$ by the similarity of triangles ACC_0 and ABB_0).

55. (a) Let l_1, l_2, l_3, l_4 be four given lines. Consider all quadrilaterals $\bar{A}\bar{B}\bar{C}\bar{D}$ similar to the given one and having the three vertices $\bar{A}$, $\bar{B}$ and $\bar{C}$ on the lines l_1, l_2 and l_3, respectively; given vertex $\bar{A}$ or the direction of side $\bar{A}\bar{B}$ of such a quadrilateral, we can construct it [see Problem 30 (a) in Section 2 and 9 (b) in Section 1]. It follows from Theorem 3 (page 71) that the vertices $\bar{D}$ of all these quadrilaterals lie on a certain line l which can easily be constructed by finding two positions of $\bar{D}$. The point of intersection of l and l_4 is the vertex D of the desired quadrilateral $ABCD$; it remains to apply the construction indicated in the solution to Problem 30 (a).

In general the problem has a unique solution; an exception occurs when $l \parallel l_4$ (then the problem has no solution), or when $l = l_4$ (the solution is undetermined).

(b) Let M_1, M_2, M_3, M_4 be the four given points. Consider all quadrilaterals $\bar{A}\bar{B}\bar{C}\bar{D}$ similar to the given one and such that the sides $\bar{A}\bar{B}$, $\bar{B}\bar{C}$ and $\bar{C}\bar{D}$ pass through the points M_1, M_2 and M_3; since the vertices $\bar{B}$ and $\bar{C}$ of these quadrilaterals lie on arcs circumscribed on M_1M_2 and M_2M_3 and subtending known angles, one can construct many such quadrilaterals. By Theorem 4 (page 72) the side $\bar{D}\bar{A}$ of each of these quadrilaterals passes through a certain point M which is easy to find by constructing two of the quadrilaterals $\bar{A}\bar{B}\bar{C}\bar{D}$. The line MM_4 will contain side DA of the desired quadrilateral. If M coincides with M_4, then the solution is undetermined.

(c) The vertex $\bar{A}$ of each quadrilateral $\bar{A}\bar{B}\bar{C}\bar{D}$ that is similar to the given one and is such that $\bar{B}\bar{C}$, $\bar{C}\bar{D}$ and $\bar{B}\bar{D}$ pass through the given points M, N and P respectively, lies, by Theorem 4, on a certain circle $\bar{S}$ (which is not difficult to construct; for this it is enough to find three positions of $\bar{A}$). Any point of intersection of $\bar{S}$ and the given circle S can serve as the vertex A of the desired quadrilateral [compare the solution to part (b)]. The problem can have two, one, or no solutions; if $\bar{S}$ and S coincide, the solution is undetermined.

56. Consider all lines $\bar{l}$ such that the ratio of the segments $\bar{A}\bar{B}$ and $\bar{B}\bar{C}$ cut off on $\bar{l}$ by the lines l_1, l_2 and l_3 has a given value; by choosing an arbitrary point A on the line l_1 we can construct $\bar{l}$ (see Problem 1 in Section 1). The points $\bar{D}$ on the lines $\bar{l}$ that are such that the segments $\bar{A}\bar{B}$, $\bar{B}\bar{C}$ and $\bar{C}\bar{D}$ have given ratios lie on a line m (see Theorem 3); this line m can be constructed easily by finding two positions of $\bar{D}$. The point of intersection of m and l_4 lies on the desired line l; it remains to apply the construction indicated in the solution to Problem 1. If $m \parallel l_4$ then the problem has no solution; if m coincides with l_4 then the solution is undetermined [compare the solution to Problem 55 (a)].

57. When the angle α is changed, $\triangle A'B'C'$ changes but remains similar to itself (and to $\triangle ABC$). Its sides always pass through certain fixed points, namely, the midpoints of the sides of ABC; therefore each of its points (and, in particular, the point of intersection of its altitudes, or of its angle bisectors, or of its medians) describes a circle. The second assertion of the problem follows from the fact that the rotation center O of all positions of triangle $A'B'C'$ (including ABC, which corresponds to the value $\alpha = 0$) coincides with the center of the circumscribed circle about $\triangle ABC$; to see this it is sufficient to note that for $\alpha = 90°$ the lines we are considering all pass through the single point O.

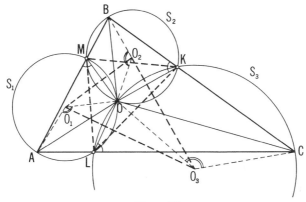

Figure 121

58. (a) If $\triangle KLM$ changes, remaining similar to itself, in such a manner that its vertices K, L and M slide on the sides BC, CA and AB of $\triangle ABC$, then all positions of $\triangle KLM$ have a common center of rotation O which lies on the circles S_1, S_2 and S_3 (see the proof of Theorem 3, in particular Figure 54a, page 72). Therefore these three circles pass through the common point O (Figure 121).

(b) Let O_1, O_2 and O_3 be the centers of circles S_1, S_2 and S_3, and let O be their common point (Figure 121). Since the quadrilateral $ALOM$ is inscribed in a circle, we have $\angle AMO + \angle ALO = 180°$ and, consequently $\angle AMO = \angle CLO$; analogously, $\angle CLO = \angle BKO$. But the equality $\angle AMO = \angle BKO = \angle CLO$ implies $\angle AO_1O = \angle BO_2O = \angle CO_3O$; therefore, triangles OO_1A, OO_2B and OO_3C are all similar to one another, and $\triangle O_1O_2O_3$ can be obtained from $\triangle ABC$ by a spiral similarity (with center O, rotation angle O_1OA, and similarity coefficient OO_1/OA).

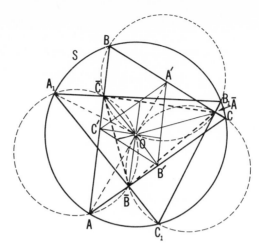

Figure 122

59. (a) Triangle $A_1B_1C_1$ is obtained from $\triangle ABC$ by a rotation about the point O through some angle α. It follows from this that $\sphericalangle A\bar{B}A_1 = \sphericalangle A\bar{C}A_1 = \sphericalangle AOA_1 = \alpha$, that is, the points A, A_1, $\bar{B}$, $\bar{C}$ and O lie on a circle (Figure 122). In the same way it can be shown that the five points B, B_1, $\bar{A}$, $\bar{C}$, O lie on a circle and so do the five points C, C_1, $\bar{A}$, $\bar{B}$, O.

Consider the triangle $A'B'C'$ formed by the mid-lines of $\triangle ABC$. We shall change this triangle, keeping it always similar to its original position (that is, similar to $\triangle ABC$), in such a manner that its vertices slide on the sides of $\triangle ABC$. All positions of the triangle have a common center of rotation—the point of intersection of the circles circumscribed about triangles $AB'C'$, $BA'C'$, $CA'B'$ [see the solution to Problem 58 (a)], that is, the point O. Assume now that one vertex of the changing triangle is $\bar{A}$; let $\bar{B}'$ and $\bar{C}'$ be its other two vertices. Then $\bar{B}'$ lies on a common circle with C, $\bar{A}$ and O (see the proof of Theorem 3); it follows that $\bar{B}' = \bar{B}$. In the same way it can be shown that vertex $\bar{C}' = \bar{C}$.

(b) The point O is a fixed point of the changing triangle $\bar{A}\bar{B}\bar{C}$; it must correspond to itself for all positions of this triangle. Since O is the point of intersection of the altitudes of $\triangle A'B'C'$ it follows that O is also the point of intersection of the altitudes of $\triangle \bar{A}\bar{B}\bar{C}$.

60. (a) l_2 is obtained from l_1 by the sum of two reflections in the sides of $\triangle ABC$; therefore the angle between l_1 and l_2 is equal to twice the angle between these sides of the triangle (see pages 50 and 30 of Volume One). Thus the angles of triangle T are determined by the angles of $\triangle ABC$ and do not depend on the position of l. (If $\triangle ABC$ were a right triangle, then two of the three lines l_1, l_2, l_3 would be parallel and therefore l_1, l_2, l_3 would not form a triangle.)

(b) Assume that the line l is rotated about some point M in the plane. Then the sides of triangle T always pass through the points M_1, M_2, M_3 that are symmetric to M with respect to the sides of $\triangle ABC$; in other words the triangle T varies in such a way that it remains similar to itself, and its sides always pass through three fixed points. In the proof of Theorem 4 (pp. 73–74) it was shown that in this case each two positions of the triangle T have the same rotation center O. If l_1 passes through O (that is, l passes through the point O' symmetric to O with respect to side AB), then, and only then, triangle T degenerates to a point (and in this case the lines l_2 and l_3 also pass through O).

Thus we see that, in general, among the lines passing through a given point M there is *only one* line l such that l_1, l_2 and l_3 meet in a point; if there were two such lines, it would mean that *all* lines through M would have this property. Now let M and N be two points and let l and $\bar{l}$ be the lines through them for which the corresponding triples of lines l_1, l_2, l_3 and $\bar{l}_1$, $\bar{l}_2$, $\bar{l}_3$, meet in a point. If H is the point of intersection of l and $\bar{l}$, then for each line through H the corresponding lines l_1, l_2, l_3 meet in a point. (The lines l and $\bar{l}$ cannot be parallel, for if l_1, l_2, and l_3 meet in a point O and if $l \parallel \bar{l}$, then the lines $\bar{l}_1$, $\bar{l}_2$, and $\bar{l}_3$ would be parallel to the corresponding lines l_1, l_2 and l_3 and would be separated from O by a distance equal to the distance between l and $\bar{l}$; therefore they cannot meet in a point.) If l passes through H then l_1 and l_2 pass through points H_1 and H_2, symmetric to H with respect to the sides of triangle ABC; since, in addition, the angle between l_1 and l_2 has a definite magnitude [see the solution to part (a)], we see that the point P of intersection of l_1 and l_2 describes a circle S (constructed on the segment H_1H_2 and subtending a known angle†).

Thus we have shown the existence of a point H such that for each line passing through H the corresponding lines l_1, l_2 and l_3 meet in a point. There cannot be two such points G and H, for otherwise through each point M there would be two lines MG and MH for each of which the corresponding triple of lines l_1, l_2, l_3 would meet in a point. To see that H is the point of intersection of the altitudes of $\triangle ABC$, and S is the circumscribed circle about this triangle, it is sufficient to note that the lines l_1, l_2, l_3, corresponding to the altitudes of $\triangle ABC$, meet at its vertices.

(c) Let l be an arbitrary line and let $\bar{l}$ be the line parallel to it and passing through H. The lines $\bar{l}_1$, $\bar{l}_2$, $\bar{l}_3$ meet in a point P; the lines l_1, l_2, and l_3 will be, as was noted in the solution to part (b), separated from the point P by a distance equal to the distance between $\bar{l}$ and l, or, what is the same thing, the distance from H to l. Thus the radius of the inscribed circle of triangle T is equal to the distance from H to l; since all the triangles T are similar to one another, it follows that the area of T depends only on the distance from H to l.

† See footnote ‡ on page 50 of Volume One.

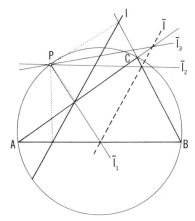

Figure 123

61. *First solution.* If the feet of the perpendiculars from P to the sides of $\triangle ABC$ lie on a line l, then the lines $\bar{l}_1$, $\bar{l}_2$ and $\bar{l}_3$ that are the reflections of $\bar{l}$ in the sides of $\triangle ABC$, meet in the point P; here $\bar{l}$ is centrally similar to l with similarity center P and similarity coefficient 2 (Figure 123). It follows that P must lie on the circumscribed circle about $\triangle ABC$ [and $\bar{l}$ must pass through the point H of intersection of the altitudes of $\triangle ABC$; see Problem 60 (b)].

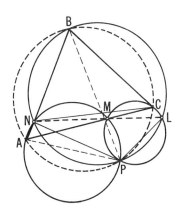

Figure 124

Second solution. Denote the feet of the perpendiculars from P onto the sides AB, BC, CA of $\triangle ABC$ by N, L, and M respectively (Figure 124). Let us show that P is the rotation center for all triangles $L'M'N'$ similar to triangle LMN, whose vertices lie on sides BC, CA, and AB of $\triangle ABC$. Indeed, from the fact that $\sphericalangle PMA = \sphericalangle PNA = 90°$, it follows that the points A, M, N and P lie on a circle, that is, P lies on circle

AMN. In the same way it can be shown that P lies on circles BNL and CLM. But the point of intersection of these circles is also the desired rotation center [see the solution to Problem 58 (a)].

Further, for example, in the case shown in Figure 124,†

$$\angle APB = \angle APN + \angle NPB = \angle AMN + \angle NLB,$$

since angles APN and AMN as well as angles NPB and NLB are inscribed in the same circle and cut off the same arcs. But

$$\angle AMN = \angle MCN + \angle MNC, \qquad \angle NLB = \angle NCB - \angle LNC$$

since

$$\angle AMN + \angle NLB = (\angle MCN + \angle NCB) + (\angle MNC - \angle LNC)$$

$$= \angle MCB + \angle MNL.$$

Comparing this equation with the preceding, we conclude that

$$\angle APB = \angle MCB + \angle MNL.$$

This equation enables us to prove the assertion of the exercise. Indeed, if the point P lies on the circle circumscribed about triangle ABC, then $\angle APB = \angle ACB$ and $\angle MNL = 0$, that is, the points M, N, L lie on a line (Figure 124). Conversely, if the points M, N and L lie on a line, then $\angle MNL = 0$ and $\angle APB = \angle ACB$; consequently, the point P lies on the circle passing through points A, B and C.

62. (a) Let l_1, l_2, l_3, l_4 be the four given lines. Circumscribe circles about the two triangles formed by lines l_1, l_2, l_3 and l_1, l_2, l_4; the point of intersection of these circles (different from the point of intersection of lines l_1 and l_2) will be denoted by P (Figure 125). From P we drop perpendiculars onto lines l_1, l_2, l_3 and l_4; let M_1, M_2, M_3 and M_4 be the feet of these perpendiculars. By the result of Problem 61 the points M_1, M_2 and M_3 lie on a line, and the points M_1, M_2 and M_4 also lie on a line; consequently, all four points lie on one line. From the same problem it follows that M_1, M_2 and M_4 also lie on a line; consequently, all four points lie on one line. From the same problem it follows that M_1, M_3 and M_4 can lie on a line only if P lies on the circle circumscribed about the triangle whose sides are formed by l_1, l_3 and l_4. In the same manner one shows that P lies on the circle circumscribed about the triangle whose sides are formed by l_2, l_3, and l_4.

(b) With the notations of Figure 58, $MP \perp AB$ (because angles MPA and MPB cut off a diameter); in the same way $MQ \perp AC$ and $ME \perp BC$. Consequently, P, Q and R are the feet of the perpendiculars dropped from the point M on the circle S circumscribed about $\triangle ABC$ onto the sides of this triangle.

† See footnote ‡ on page 50 of Volume One.

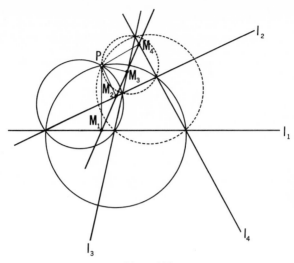

Figure 125

(c) Let $ABCD$ be a quadrilateral inscribed in a circle, and let $AB = a$, $BC = b$, $CD = c$, $DA = d$, $AC = e$, $BD = f$ (Figure 126). From D drop perpendiculars DR, DS and DT to sides AB, BC and CA, respectively, of $\triangle ABC$; by the result of the preceding problem the feet R, S and T of these perpendiculars are collinear. It is clearly possible to circumscribe a circle about the quadrilateral $ATDR$; the segment AD will be a diameter of this circle. It follows from this that

$$TR = AD \sin \sphericalangle TDR = d \sin \sphericalangle TDR.$$

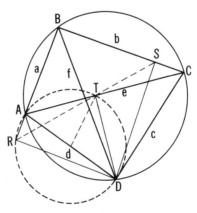

Figure 126

But $\angle TDR = \angle BAC$ (angles with perpendicular sides), and it follows from considering $\triangle ABC$ that $\sin \angle BAC = BC/2r = b/2r$, where r is the radius of the circle circumscribed about the quadrilateral $ABCD$. Thus we obtain

$$TR = d \frac{b}{2r} = \frac{bd}{2r}.$$

In an entirely analogous manner we obtain the relations

$$TS = \frac{ac}{2r}, \qquad RS = \frac{ef}{2r}.$$

Further, since the points R, S and T lie on a line (see Figure 126), we have

$$RT + TS = RS$$

or

$$\frac{bd}{2r} + \frac{ac}{2r} = \frac{ef}{2r}.$$

Multiplying both sides of this equation by $2r$ we have

$$bd + ac = ef,$$

which was to be proved.

63. The circles circumscribed about the four triangles formed by our four lines meet in a common point P [Problems 35, 62 (a)]; the feet of the perpendiculars dropped from P onto the lines l_1, l_2, l_3 and l_4 lie on a line m (Problem 61). The points H_1, H_2, H_3 and H_4 of intersection of the altitudes of the four triangles under consideration are centrally similar to the points of intersection of the lines PH_1, PH_2, PH_3 and PH_4 with the line m, with similarity center P and similarity coefficient 2 (see the first solution to Problem 61); consequently they also lie on a line, parallel to m and passing through the point P' that is symmetric to P with respect to m.

64. Denote the points of intersection of circles S_1 and S_2 by the letters A and B; further, denote by P and Q the points of intersection of line MB with the circles S_1 and S_2, respectively (Figure 127). Since we are given the ratio m/n of the lengths of the segments of the tangents from the point M to the given circles, we can also consider as known the ratio

$$\frac{MP}{MQ} = \frac{MP}{MQ} \cdot \frac{MB}{MB} = \frac{m^2}{n^2}.$$

From this equation we see that if $m = n$ the points P and Q coincide, and M is on AB; therefore in what follows we shall assume that $m \neq n$.

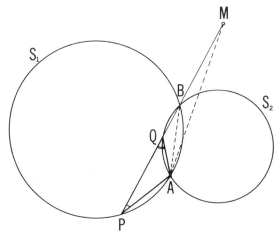

Figure 127

Consider now $\triangle AQP$. It changes when M is moved; however, the angle APQ remains constant since it always cuts off a constant arc on the circle S_1, and $\measuredangle AQP = 180° - \measuredangle AQB$ also remains constant since $\measuredangle AQB$ cuts off a constant arc on circle S_2. Consequently, the triangle remains similar to itself. Vertex A of the triangle does not move, and P describes the circle S_1; consequently the point M (lying on side QP of the triangle and such that $MP/MQ = m^2/n^2$) describes a circle S that is obtained from S_1 by a spiral similarity with center A, rotation angle PAM, and similarity coefficient AM/AP.

Clearly the circle S passes through the point A; we now show that it also passes through B. For this it is sufficient to find a point C on the circle S_1 such that $\measuredangle CAB = \measuredangle PAM$; since $\measuredangle ACB = \measuredangle APM$, the triangles ACB and APM will be similar $(AC/AB = AP/AM)$ and, therefore, the spiral similarity considered above will carry C into B.

The desired locus can now be constructed as follows. Through one of the points A and B of intersection of the circles S_1 and S_2 pass an arbitrary line meeting S_1 and S_2 in points P and Q. Construct a point M on this line such that $MP/MQ = m^2/n^2$, where m/n is the given ratio of the lengths of the tangents. The circle through A, B and M (more precisely, the arc of this circle that lies outside S_1 and S_2) is the desired locus.

65. All such triangles $A'B'C'$ have, with $\triangle ABC$, a common rotation center O coinciding with the point of intersection of the circles $AB'C'$ and $BA'C'$ (see the proof of Theorem 3). It only remains to consider the particular position of $\triangle A'B'C'$ in which A', B', and C' are the midpoints of the sides of $\triangle ABC$.

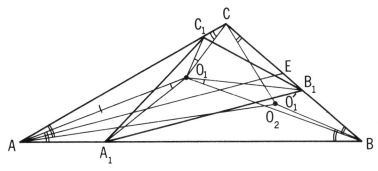

Figure 128

66. (a) Since O_1 is the rotation center of triangles ABC and $A_1B_1C_1$, we have (see Figure 128)

$$\angle AO_1A_1 = \angle BO_1B_1 = \angle CO_1C_1, \qquad \frac{O_1A}{O_1A_1} = \frac{O_1B}{O_1B_1} = \frac{O_1C}{O_1C_1} \qquad (*)$$

and, therefore, triangles AO_1A_1, BO_1B_1, CO_1C_1 are all similar to one another. From this we see that

$$\angle O_1AB = \angle O_1BC = \angle O_1CA. \qquad (**)$$

In the same way it can be shown that

$$\angle O_2BA = \angle O_2CB = \angle O_2AC. \qquad (***)$$

Conversely, if the point O_1 is such that condition (**) is satisfied for it, then it is the first rotation center of $\triangle ABC$. Indeed, pass three lines O_1A_1, O_1B_1, O_1C_1 through O_1 meeting the sides of $\triangle ABC$ in points A_1, B_1, C_1 and forming equal angles with these sides. Triangles AO_1A_1, BO_1B_1 and CO_1C_1 will be similar (by equality of angles); therefore condition (*) is fulfilled and O_1 is the rotation center of triangles ABC and $A_1B_1C_1$, that is, O_1 is the first rotation center of $\triangle ABC$. In the same way it can be shown that a point O_2 is the second rotation center of $\triangle ABC$ if condition (***) is satisfied.

We now prove that $\angle O_1AB = \angle O_2AC$. For this purpose we construct the lines symmetric to lines AO_1, BO_1 and CO_1 with respect to the bisectors of the corresponding angles, and we show that these three lines meet in a common point O_1'. Denote the distance from O_1 to the sides AB, BC, and CA of $\triangle ABC$ by m, n, and p. The line AO_1 is the locus of points whose distances from sides AB and AC of the triangle has the ratio $p:m$ (Figure 128). In the same way the line symmetric to BO_1 with respect to the bisector of angle B is the locus of points whose distances from BA and AC are in the ratio $m:n$, and the line symmetric to CO_1 with respect to the bisector of angle C is the locus of points whose distances from CA and CB are in the ratio $p:n$. From this it follows that the distances from the point O_1' of intersection of the last two lines to sides AB and AC is in the ratio $m:p$, that is, O_1' also belongs to the first of the three lines.

From condition (**), which the point O_1 satisfies, it follows that O_1' satisfies the relation

$$\angle O_1'BA = \angle O_1'CB = \angle O_1'AC$$

[since $\angle O_1AB = \angle O_1'AC$, $\angle O_1BC = \angle O_1'BA$, $\angle O_1CA = \angle O_1'CB$]. Therefore O_1' coincides with the second rotation center O_2 of triangle ABC and, therefore, $\angle O_1AB = \angle O_2AC$.

(b) Let the point O be both the first and second rotation center of $\triangle ABC$ (Figure 129). Since O is the first rotation center, we have $\angle OAB = \angle OBC = \angle OCA$; since O is also the second rotation center, we also have $\angle OBA = \angle OCB = \angle OAC$. But it now follows at once that $\angle A = \angle B = \angle C$, that is, $\triangle ABC$ is equilateral.

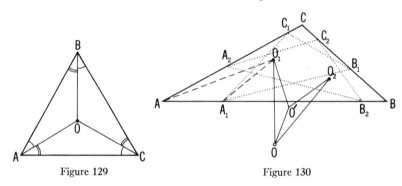

Figure 129   Figure 130

(c) Here it is most convenient to use the results of Problem 68 below. By these results the centers O_1' and O_2' of the circles circumscribed about the congruent triangles $A_1B_1C_1$ and $A_2B_2C_2$ coincide (in Figure 130 these centers coincide with O'). Since $\triangle A_1B_1C_1$ is obtained from $\triangle ABC$ by a spiral similarity with center O_1, rotation angle α, and some coefficient k, we have $\angle O'O_1O = \alpha$, $O_1O'/O_1O = k$; since $A_2B_2C_2$ is obtained from ABC by a spiral similarity with center O_2, the same rotation angle α (because A_1B_1 and A_2B_2 form equal angles with AB), and the same similarity coefficient k (because triangles $A_1B_1C_1$ and $A_2B_2C_2$ are congruent), we have $\angle O'O_2O = \alpha$, $O_2O'/O_2O = k$. It follows from this that triangles O_1OO' and O_2OO' are similar, and since they have a common side OO', they are congruent; therefore $OO_1 = OO_2$.

Note that from the similarity of the triangles O_1OO' and O_1AA_1 it follows that $\angle O_1OO' = \angle O_1AA_1$ and, therefore, $\angle O_1OO_2 = 2\phi$, where ϕ is the common value of the angles O_1AB, O_1BC, O_1CA, O_2AC, O_2CB, O_2BA. From this and from the result of part (d) it follows that $O_1O_2 \le O_1O = O_2O$ (and the equality sign holds only for equilateral triangles, when all three points O_1, O_2 and O coincide).

(d) For the proof of this theorem we shall use the construction considered in Problem 70 (see Figure 137, p. 162). Let us determine the product $AO_1 \cdot O_1A' = BO_1 \cdot O_1B' = CO_1 \cdot O_1C'$ in terms of the radius R of the circumscribed circle and the angle ϕ. From the similarity of triangles AO_1C' and $A'BO_1$ [see Problem 70 (b)] it is easy to obtain

$$\frac{AO_1}{AC'} = \frac{A'B}{A'O_1}$$

and, therefore,

$$AO_1 \cdot O_1A' = AC' \cdot A'B.$$

But since arc $AC' = $ arc $A'B = 2\phi$, we have $AC' = A'B = 2R \sin \phi$. Thus $AO_1 \cdot O_1A' = 4R^2 \sin^2 \phi$.

But, on the other hand (see Figure 137),

$$AO_1 \cdot O_1A' = MO_1 \cdot O_1N = (R - OO_1)(R + OO_1)$$

$$= R^2 - OO_1{}^2 \leq R^2.$$

From this it follows that

$$4R^2 \sin^2 \phi \leq R^2; \qquad \sin^2 \phi \leq \tfrac{1}{4}, \qquad \phi \leq 30°,$$

while $\phi = 30°$ only if O_1 coincides with O $(OO_1 = O)$ and $\triangle ABC$ is equilateral [see parts (c), (b)].

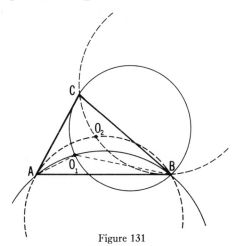

Figure 131

67. In order that O_1 be the first rotation center of $\triangle ABC$ it is necessary and sufficient that

$$\measuredangle O_1AB = \measuredangle O_1BC = \measuredangle O_1CA.$$

Assume that the first rotation center has been found, and construct on the segment AB the arc of the circle passing through O_1 (Figure 131).

Since $\sphericalangle O_1AB = \sphericalangle O_1BC$, it follows that $\sphericalangle O_1BC$ is equal to one half the measure of arc BO_1 of the circle; therefore the line BC is tangent to the circle BO_1A. In the same way it can be shown that the sides CA and AB of the triangle are respectively tangent to the circle passing through the points B, C and O_1 and the circle through the points C, A and O_1. Therefore the first rotation center O_1 of triangle ABC can be constructed as the point of intersection of two circles: one circle passing through A and B and tangent to side BC and the other passing through B and C and tangent to CA. The second rotation center can be constructed in an analogous manner.

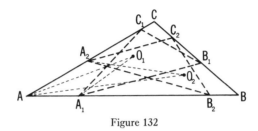

Figure 132

68. (a) Triangle $A_1B_1C_1$ can be obtained from $\triangle ABC$ by means of a rotation about O_1 through angle AO_1A_1, followed by a central similarity with coefficient O_1A_1/O_1A; therefore the angle between lines AB and A_1B_1 is equal to angle AO_1A_1 (Figure 132). In the same way it can be shown that the angle between lines AB and A_2B_2 is equal to angle AO_2A_2. Therefore from the conditions of the problem we see that

$$\sphericalangle AO_1A_1 = \sphericalangle AO_2A_2 .$$

But it follows from Problem 66 (a) that $\sphericalangle O_1AA_1 = \sphericalangle O_2AA_2$. Thus triangles AO_1A_1 and AO_2A_2 are similar, and therefore

$$\frac{O_1A_1}{O_1A} = \frac{O_2A_2}{O_2A},$$

that is, the similarity coefficient of triangles ABC and $A_1B_1C_1$ is equal to the similarity coefficient of triangles ABC and $A_2B_2C_2$. It follows from this that triangles $A_1B_1C_1$ and $A_2B_2C_2$ are congruent.

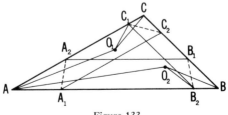

Figure 133

(b) Let us prove that $B_2C_1 \parallel BC$ (Figure 133). From the solutions to Problems 66 (a), 68 (a) it follows that triangles CO_1C_1 and BO_2B_2 are similar, and therefore

$$\frac{CC_1}{BB_2} = \frac{CO_1}{BO_2}.$$

Further, from Problem 66 (a) it follows that $\angle O_1CA = \angle O_2BA$ and $\angle O_1AC = \angle O_2AB$; therefore triangles CO_1A and BO_2A are similar and

$$\frac{CO_1}{BO_2} = \frac{AC}{AB}.$$

From this we have

$$\frac{CC_1}{BB_2} = \frac{AC}{AB},$$

which proves our assertion.

In exactly the same way one proves that $C_2A_1 \parallel CA_1$, and that $A_2B_1 \parallel AB$.

Let us prove further that the line A_1A_2 is anti-parallel to side BC of $\triangle ABC$. From the similarity of triangles AO_1A_1 and AO_2A_2 we have

$$\frac{AA_1}{AA_2} = \frac{O_1A}{O_2A}.$$

From the similarity of triangles CO_1A and BO_2A we have

$$\frac{O_1A}{O_2A} = \frac{AC}{AB}.$$

Comparing the two proportions just obtained we find that

$$\frac{AA_1}{AA_2} = \frac{AC}{AB},$$

from which it follows that triangles AA_1A_2 and ACB are similar, and consequently, that lines A_1A_2 and BC are anti-parallel. In the same way it can be shown that B_1B_2 is anti-parallel to CA and that C_1C_2 is anti-parallel to AB.

(c) Consider the quadrilateral $B_1C_1A_1B_2$. In this quadrilateral $\angle B_1C_1A_1 = \angle C$, since triangles $A_1B_1C_1$ and ABC are similar; in addition, $\angle A_1B_2B_1 = 180° - \angle B_1B_2B$ (see Figure 134). But since B_1B_2 is anti-parallel to side CA of $\triangle ABC$, we have $\angle B_1B_2B = \angle C$ and, consequently,

$$\angle B_1C_1A_1 + \angle A_1B_2B_1 = 180°.$$

From this we see that B_2 lies on the circle circumscribed about $\triangle A_1B_1C_1$. In the same way it can be shown that C_2 and A_2 lie on this circle.

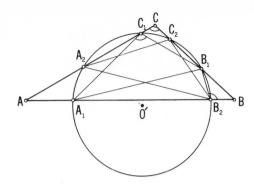

Figure 134

69. (a) Since in the right triangles AA_1O_1, BB_1O_1 and CC_1O_1 the angles O_1AA_1, O_1BB_1 and O_1CC_1 are all equal [see Problem 66 (a)], these triangles are all similar (Figure 135). Therefore, $\angle AO_1A_1 = \angle BO_1B_1 = \angle CO_1C_1$ and $O_1A/O_1A_1 = O_1B/O_1B_1 = O_1C/O_1C_1$; that is, $\triangle A_1B_1C_1$ can be obtained from $\triangle ABC$ by a spiral similarity with center O_1. Thus, $\triangle A_1B_1C_1 \sim \triangle ABC$. It can be shown in the same way that $\triangle A_2B_2C_2 \sim \triangle ABC$.

The right triangles AA_1O_1 and AA_2O_2 are similar [see Problem 66 (a)]; consequently angles AO_1A_1 and AO_2A_2, through which triangles $A_1B_1C_1$ and $A_2B_2C_2$, respectively, were rotated with respect to $\triangle ABC$, are equal. Therefore the angles formed by the lines A_1B_1 and A_2B_2 with AB are also equal, and triangles $A_1B_1C_1$ and $A_2B_2C_2$, inscribed in $\triangle ABC$, satisfy the conditions of Problem 68; hence all results of that problem can be applied to them.

From the similarity of triangles O_1OO' and O_1AA_1, where O_1OO' is the triangle considered in the solution to Problem 66 (a) (see Figure 135) we have $\angle O_1O'O = \angle O_1A_1A = 90°$. From this it follows that the point O', the common center of the circumscribed circles about triangles $A_1B_1C_1$ and $A_2B_2C_2$, is the midpoint of O_1O_2.

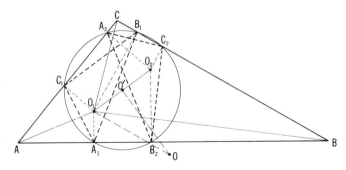

Figure 135

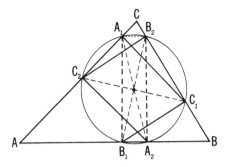

Figure 136

(b) If the vertices of $\triangle A_1B_1C_1$, whose sides are parallel to the altitudes of $\triangle ABC$, lie on the sides of ABC in the order pictured in Figure 136, then A_1B_1 can be parallel to the altitude CF or to the altitude BE (but not to the altitude AD!). If $A_1B_1 \parallel CF$, then $A_1C_1 \parallel BE$; if $A_1B_1 \parallel BE$, then $B_1C_1 \parallel CF$. Thus there are two possibilities for inscribing in the given triangle ABC a triangle whose sides are parallel to the altitudes of ABC, that is, there are two such inscribed triangles [see Problem 9 (b)]. In Figure 136 these triangles are denoted by $A_1B_1C_1$ and $A_2B_2C_2$. They are similar to $\triangle ABC$ (we use the same letters to indicate corresponding vertices). It is easy to see that all the conditions of Problem 68 are satisfied for triangles $A_1B_1C_1$ and $A_2B_2C_2$; hence they are congruent [see Problem 68 (a)].

Now consider the quadrilateral $A_1B_1A_2B_2$. We have $A_1B_1 = A_2B_2$, $A_1B_1 \parallel A_2B_2$, and $A_1B_1 \perp B_1A_2$. Consequently this quadrilateral is a rectangle and its diagonals are equal and are bisected by their point of intersection. In the same way it can be shown that the segments B_1B_2 and C_1C_2 are equal and bisected by their point of intersection. Therefore all three segments joining corresponding vertices of triangles $A_1B_1C_1$ and $A_2B_2C_2$ are equal and are bisected by their common point of intersection. This point is the center of the circle circumscribed simultaneously about triangles $A_1B_1C_1$ and $A_2B_2C_2$ [see Problem 68 (c)].

70. (a) Denote the value of $\sphericalangle A'AB$ by ϕ. Since

$$\sphericalangle A'AB = \sphericalangle B'BC = \sphericalangle C'CA = \phi,$$

we have $\operatorname{arc} BA' = \operatorname{arc} CB' = \operatorname{arc} AC' = 2\phi$ and triangle $C'A'B'$ is obtained from $\triangle ABC$ by a rotation through an angle of 2ϕ about the center O of the circumscribed circle (Figure 137).

(b) For example, for triangle AO_1C' we have

$$\sphericalangle AO_1C' = \frac{\operatorname{arc} AC' + \operatorname{arc} CA'}{2} = \frac{\operatorname{arc} BA' + \operatorname{arc} CA'}{2} = \frac{\operatorname{arc} BC}{2} = \sphericalangle BAC,$$

$$\measuredangle C'AO_1 = \frac{\operatorname{arc} C'B + \operatorname{arc} BA'}{2} = \frac{\operatorname{arc} C'B + \operatorname{arc} AC'}{2} = \frac{\operatorname{arc} AB}{2} = \measuredangle ACB,$$

and therefore $\triangle AO_1C'$ is similar to $\triangle ABC$. The proof is analogous for the five remaining triangles.

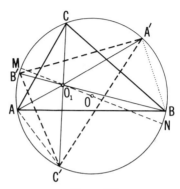

Figure 137

71. Join the first rotation center O_1 of $\triangle ABC$ to all the vertices of the triangle (Figure 138). The point M lies in one of the triangles ABO_1, BCO_1, CAO_1. If, for example, M lies inside (or on the boundary) of $\triangle ABO_1$ (Figure 138), then $\measuredangle MAB \leq \measuredangle O_1AB \leq 30°$ [see Problem 66 (d)]. Similarly one shows that at least one of the angles MAC, MBC, MBA is less than or equal to 30° (along with triangles ABO_1, BCO_1, CAO_1 it is necessary to consider triangles ACO_2, CBO_2, BAO_2, where O_2 is the second rotation center of the triangle).

In fact, it follows from the proof that at least one of the angles MAB, MBC, MCA is always strictly less than 30°, with the unique exception of the case when $\triangle ABC$ is equilateral and M is its center (in this case $\measuredangle MAB = \measuredangle MBC = \measuredangle MCA = 30°$).

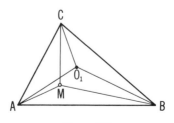

Figure 138

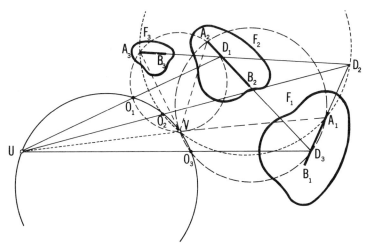

Figure 139

72. (a), (b). Let V be a point of intersection, different from A_1, of the circles circumscribed about triangles $A_1A_2D_3$ and $A_1A_3D_2$ (Figure 139). In this case, denoting the angles of $\triangle D_1D_2D_3$ by D_1, D_2 and D_3, we have (Figure 139†)

$$\angle A_1VA_2 = \angle D_3, \qquad \angle A_1VA_3 = 180° - \angle D_2$$

and, consequently,

$$\angle A_2VA_3 = \angle A_3VA_1 - \angle A_2VA_1 = 180° - \angle D_2 - \angle D_3 = \angle D_1$$

and, therefore, the circle circumscribed about $\triangle A_2A_3D_1$ also passes through V.

Note that, by the construction of the rotation center (see page 43, Figure 30b), O_1 lies on the circle circumscribed about $\triangle A_2A_3D_1$, O_2 lies on the circle circumscribed about $\triangle A_1A_3D_2$, and O_3 lies on the circle circumscribed about $\triangle A_1A_2D_3$; moreover, as we proved, all these circles pass through the common point V.

Now denote the point of intersection of the lines D_2O_2 and D_3O_3 by U. Then we have (see Figure 139†)

$$\angle O_2VO_3 = \angle D_2O_2V + \angle D_3O_3V - \angle O_2UO_3$$

(for the proof it is enough to divide the quadrilateral O_2VO_3U into two triangles by the diagonal UV and to apply the theorem of the exterior angles to each of these triangles). Because the points D_2, O_2, V and A_1 lie on a circle we have

$$\angle D_2O_2V = 180° - \angle D_2A_1V$$

† See footnote ‡ on page 50 of Volume One.

and similarly

$$\sphericalangle D_3O_3V = 180° - \sphericalangle D_3A_1V,$$

from which it follows that

$$\sphericalangle D_2O_2V + \sphericalangle D_3O_3V = 180°$$

and, therefore,

$$\sphericalangle O_2VO_3 = 180° - \sphericalangle O_2UO_3,$$

that is, the points O_2, O_3, U and V lie on a circle.

Let $\bar{V}$ denote the common point of intersection of the circles circumscribed about triangles $B_1B_2D_3$, $B_1B_3D_2$, and $B_2B_3D_1$. In the same manner as in the preceding paragraph we show that O_2, O_3, U and $\bar{V}$ lie on a common circle. Thus we see that the five points O_2, O_3, V, $\bar{V}$ and U lie on a circle. In the same way it can be shown that the points O_1, O_3, V, $\bar{V}$ lie on a circle. But from the fact that the two quadruples of points O_1, O_3, V, $\bar{V}$ and O_2, O_3, V, $\bar{V}$ each separately lie on a circle we see that these two circles must coincide with one another and with the similarity circle of the figures F_1, F_2, F_3; this completes the proof of the proposition in part (b). Further, we see that the point U of intersection of D_2O_2 and D_3O_3 lies on this circle. In other words, the line D_2O_2 passes through the point U (distinct from O_3) of intersection of the line D_3O_3 with the similarity circle of the figures F_1, F_2, F_3; in the same way it can be shown that the line D_1O_1 passes through the same point, which completes the proof of the proposition of part (a).

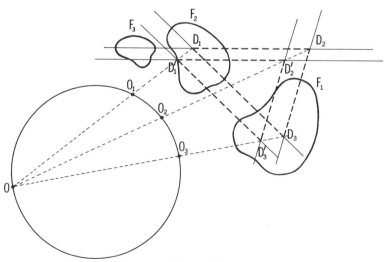

Figure 140

(c) Assume first that the sides of $\triangle D_1'D_2'D_3'$ are parallel to the corresponding sides of $\triangle D_1D_2D_3$ (Figure 140). From the fact that O_3 is the rotation center of figures F_1 and F_2 it follows easily that the distances of the lines $D_2'D_3'$ and $D_1'D_3'$ from O_3 are proportional to the distances of D_2D_3 and D_1D_3 from O_3. It follows from this that D_3' lies on the line O_3D_3. In the same way it can be shown that D_2' lies on O_2D_2 and that D_1' lies on O_1D_1. Taking into account the result of part (a) we see that the lines D_1D_1', D_2D_2' and D_3D_3' meet in a common point O which is clearly the similarity center of the triangles $D_1D_2D_3$ and $D_1'D_2'D_3'$; this point lies on the similarity circle of the figures F_1, F_2 and F_3.

Now assume that the sides of $\triangle D_1'D_2'D_3'$ are not parallel to the corresponding sides of $\triangle D_1D_2D_3$. Let A_1B_1 and A_1C_1, A_2B_2 and A_2C_2, A_3B_3 and A_3C_3 be three pairs of corresponding segments of the figures F_1, F_2, and F_3; let $D_1D_2D_3$ and $D_1'D_2'D_3'$ be the triangles whose sides are formed by the lines A_1B_1, A_2B_2 and A_3B_3, and the lines A_1C_1, A_2C_2 and A_3C_3, respectively (see Figure 63c in the text). The triangles $D_1D_2D_3$ and $D_1'D_2'D_3'$ are similar since the angle between lines A_1B_1 and A_1C_1 of figure F_1 is equal to the angle between the corresponding lines A_2B_2 and A_2C_2 of figure F_2 and to the angle between A_3B_3 and A_3C_3 of figure F_3. The rotation center of triangles $D_1D_2D_3$ and $D_1'D_2'D_3'$ lies on the circle passing through the three points D_3, D_3' and A_1 (see the construction of the rotation center of two figures on page 43). But since $\sphericalangle D_3A_1D_3' = \sphericalangle D_3A_2D_3'$, this circle also passes through A_2. Thus the rotation center of triangles $D_1D_2D_3$ and $D_1'D_2'D_3'$ lies on the circle circumscribed about $\triangle A_1A_2D_3$. In a similar manner the rotation center lies on the circles circumscribed about triangles $A_1A_3D_2$ and $A_2A_3D_1$. By part (b) the point of intersection of these circles lies on the similarity circle of the figures F_1, F_2, F_3 (compare Figure 63c and 63b).

73. (a) Let l_1, l_2, l_3 be three corresponding lines of the figures F_1, F_2, and F_3 meeting in a point W; let l_1' be a line in the figure F_1 parallel to l_1, let l_2', l_3' be the lines corresponding to l_1' in figures F_2 and F_3, and let $D_1D_2D_3$ be the triangle whose sides are formed by the lines l_1', l_2', l_3' (Figure 141). Clearly $l_2' \parallel l_2$ and $l_3' \parallel l_3$. Further, since O_3 is the rotation center of F_1 and F_2, the distance from O_3 to l_1 and l_2 is proportional to the distance from O_3 to l_1' and l_2'. It follows from this that the line O_3D_3 passes through W. In the same way it can be shown that lines O_2D_2 and O_1D_1 pass through W. After this it only remains to apply the result of Problem 72 (a).

(b) Note first of all that angle D_1 of triangle $D_1D_2D_3$ (Figure 141) does not depend on the choice of the lines l_1', l_2', l_3'; this is the angle of rotation of the spiral similarity carrying F_2 into F_3. Further, the ratio of the distances from the point O_1 to the lines l_2' and l_3' does not depend on the choice of these lines; it is equal to the similarity coefficient of figures F_2 and F_3. It follows from this that the angles $O_1D_1D_2$ and $O_1D_1D_3$ have constant values. If now J_2 and J_3 are the points of intersection of l_2 and l_3 with the similarity circle of F_1, F_2 and F_3, then $\angle O_1WJ_2 = \angle O_1D_1D_3$ and $\angle O_1WJ_3 = \angle O_1D_1D_2$; the arcs O_1J_2 and O_1J_3 have constant values and, consequently, J_2 and J_3 do not depend on the choice of the lines l_1, l_2 and l_3. In the same way l_1 also meets the similarity circle of figures F_1, F_2 and F_3 in the fixed point J_1.

We leave it to the reader to prove for himself that, conversely, if U is an arbitrary point on the similarity circle of F_1, F_2 and F_3, then the lines UJ_1, UJ_2 and UJ_3 are corresponding lines of these figures.

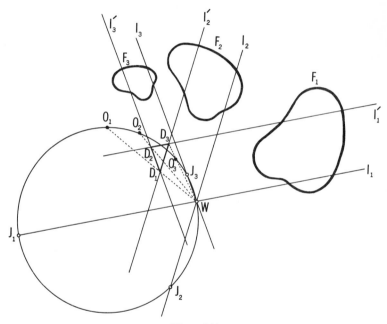

Figure 141

74. (a) Let B' be the point symmetric to B with respect to the line l (Figure 142a). If X' is an arbitrary point on l, then

$$AX' + X'B = AX' + X'B'.$$

Therefore, the sum $AX' + BX'$ will be smallest when the sum $AX' + B'X'$ is smallest, that is, when X' coincides with the point X of intersection of AB' with l.

(b) Let B' be symmetric to B with respect to the line l (Figure 142b). If X' is an arbitrary point on l, then $AX' - B'X' \leq AB'$. And since $AX' - BX' = AX' - B'X'$, the difference $AX' - BX'$ will be minimum when X' coincides with the point X of intersection of AB' with l.

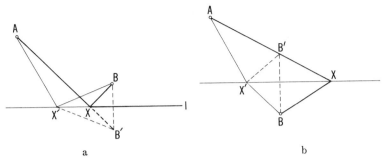

a   b

Figure 142

75. (a) *First solution.* Let PXY be an arbitrary triangle inscribed in $\triangle ABC$, one of whose vertices is P. Reflect $\triangle ABC$ together with $\triangle PXY$ in the line BC; the triangle $A'BC$ thus obtained, together with the triangle $P'XY'$ inscribed in it, we reflect in the line CA' (Figure 143). Since, with the notations of Figure 143, $XY = XY'$ and $YP = YP''$, the perimeter of $\triangle PXY$ is equal to the length of the polygonal line $PXY'P''$.

Two cases are possible now. If the segment PP'' crosses the line BC between the points B and C (and therefore cuts the line CA' between the points C and A'), then it will be shorter than any polygonal line $PXY'P''$ and the desired triangle has been found (triangle PMN in Figure 143; M is the point where PP'' crosses BC, N is the point symmetric to the point N' with respect to the line BC, and N' is the point of intersection of PP'' with CA'). If the segment PP'' crosses the line BC outside the interval BC, then the shortest polygonal line $PXY'P''$ will be the one for which X and Y' coincide with C. In that case the desired triangle degenerates to the segment PC described twice.

It remains to explain when each of these two cases occurs. For this we note that $\triangle A'B''C$ is obtained from $\triangle ABC$ by a rotation about C through an angle equal to twice angle C (because $A'B''C$ is obtained from ABC as the result of two successive reflections in the lines BC and CA' which form the angle C; see Proposition 3 in Section 1, Chapter II, page 50 of Volume One); therefore $\sphericalangle PCP'' = 2\sphericalangle C$. From this it follows at once that if $\sphericalangle C < 90°$, then the line PP'' crosses side BC of the triangle, while if $\sphericalangle C \geq 90°$, then PP'' meets the line BC either at C, or at a point lying on the continuation of BC past C.

Second solution. Once more let PXY be an arbitrary triangle inscribed in $\triangle ABC$; let P' and P'' be the points symmetric to P with respect to BC and CA (Figure 144). Since $PX = P'X$ and $PY = P''Y$, the perimeter of $\triangle PXY$ is equal to the length of the polygonal line $P'XYP''$. Therefore, if $P'P''$ crosses the two sides AC and BC of $\triangle ABC$ in points M and N, then $\triangle PMN$ is the desired triangle. If $P'P''$ does not cross the segments AC and BC, then the desired triangle degenerates to the segment PC

described twice. In a manner similar to the first solution it can be shown that the first case holds if angle C of the triangle is less than 90°, and the second if $\angle C \geq 90°$.

We note that in essence the second solution is not very different from the first (compare Figures 143 and 144).

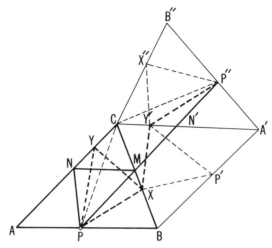

Figure 143

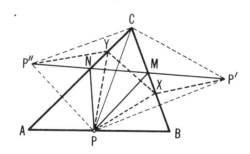

Figure 144

(b) *First solution.* We shall assume that the angle at vertex C in the given triangle is an acute angle. Let P be an arbitrary point on side AB; using the first solution to part (a) we inscribe in ABC a triangle PMN having the smallest possible perimeter, equal to the length of the segment PP'' (see Figure 143). It only remains now to choose the point P in such a way that the segment PP'' will be as small as possible. Recall that $\angle PCP'' = 2\angle C$, that is, it does not depend on the choice of the point P; therefore the base PP'' of the isosceles triangle PCP'' with the given vertex angle $2 \angle C$ will be a minimum if the side CP is as small as possible. At this point we must consider two cases separately.

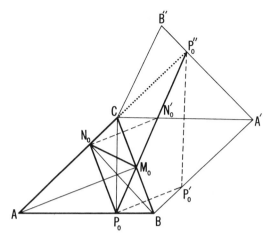

Figure 145

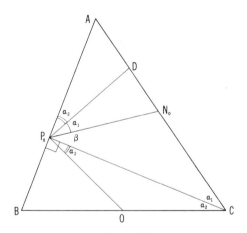

Figure 146

1°. *The angles at vertices A and B of triangle ABC are both acute* (we have an acute triangle). In this case the segment CP has the shortest possible length when the point P is the foot P_0 of the altitude CP_0 of triangle ABC (Figure 145). It is easy to show that the vertices M_0 and N_0 of triangle $P_0M_0N_0$, obtained from this choice of P_0, are also the feet of the altitudes of triangle ABC. Indeed, from Figure 145 it follows that

$$\angle N_0 P_0 A = \angle C P_0 A - \angle C P_0 N_0$$

$$= 90° - \angle C P_0'' N_0' = 90° - \frac{180° - 2 \angle C}{2} \ ;$$

that is, a circle can be circumscribed about quadrilateral BCN_0P_0 and $\sphericalangle BN_0C = \sphericalangle CP_0B = 90°$. In the same way it can be shown that $AM_0 \perp BC$.^{T}

2°. If, for example, A *is a right angle or an obtuse angle*, then segment CP will be smallest when P coincides with vertex A. In this case the desired triangle degenerates to the altitude AM_0 described twice.

Second solution. In solving part (b) one can also start from the second solution to part (a). Since the perimeter of $\triangle MNP$ (see Figure 144) is equal to $P'P''$, and $CP' = CP'' = CP$ and $P'CP'' = 2 \sphericalangle C$, the problem is reduced to finding a point P for which CP has the smallest possible value. For the rest, see the first solution.

76. We begin by solving a problem analogous to that in Problem 75 (a), namely: assume that a point P is given on side AD of quadrilateral $ABCD$, and let us try to find the quadrilateral of minimum perimeter inscribed in $ABCD$ and with one vertex at P. This problem admits a solution rather like the first solution to Problem 75 (a). Reflect the quadrilateral $ABCD$ in side AB; then reflect the quadrilateral $ABC'D'$ obtained in this manner in side BC'; finally, reflect the quadrilateral $A'BC'D''$ obtained in this manner in side $C'D''$, to obtain quadrilateral $C'D''A''B'$ (Figure 147a).

---

^{T} The calculation presented above has used the facts that in triangle CP_0P_0'', $\sphericalangle P_0CP_0'' = 2 \sphericalangle C$ and $\sphericalangle CP_0P_0'' = \sphericalangle CP_0''P_0 = \sphericalangle CP_0''N_0$, and so $2 \sphericalangle CP_0''N_0 = 180° - 2 \sphericalangle C$.

To see that a circle can be circumscribed about quadrilateral BCN_0P_0, given that $\sphericalangle N_0P_0A = \sphericalangle C$, we may proceed as follows. The segment CP_0 divides $\sphericalangle C$ into two parts. Let $\alpha_1 = \sphericalangle P_0CN_0$ and let $\alpha_2 = \sphericalangle BCP_0$. Draw a line through P_0 making angle α_1 with P_0N_0 and let this line meet AN_0 at the point D. Thus $\sphericalangle DP_0A = \alpha_2$ (see Figure 146). Let $\beta = \sphericalangle CP_0N_0$; thus $\alpha_1 + \alpha_2 + \beta = 90°$. Let O be the midpoint of BC. Since $\sphericalangle CP_0B = 90°$, the circle circumscribed about triangle CP_0B will have BC as a diameter, and hence O will be the center of this circle. Then $OC = OP_0$ since they are radii of this circle, and hence $\sphericalangle OP_0C = \sphericalangle OCP_0 = \alpha_2$.

Now consider the circle S circumscribed about $\triangle CP_0N_0$. Our task is to show that S has its center at O. Since $\sphericalangle P_0CN_0 = \sphericalangle DP_0N_0$, it follows that P_0D is tangent to S. Since $\alpha_1 + \alpha_2 + \beta = 90°$ it follows that P_0O runs along the radius to S at P_0. Finally, since $OP_0 = OC$ and C lies on S, we see that O must be the center of S, as was to be proved.

The above calculations were designed to show that when P_0 is chosen as described in the solution to the problem, then automatically M_0 and N_0 are the feet of perpendiculars. This can also be proved as follows. Suppose, for example, that N_0 is not the foot of the perpendicular from B onto side AC. Then by a discussion exactly like the discussion concerning P_0 we see that we can obtain an inscribed triangle of smaller perimeter by choosing the vertex N to lie at the foot of the perpendicular and then constructing the other two vertices P and M as in the solution to part (a). But this is impossible since we have already seen that the best choice for P is at the point P_0, which is the foot of the perpendicular from C.

Suppose that in this process an arbitrary inscribed quadrilateral $PXYZ$ passes successively through the positions $P'XY'Z'$, $P''X'Y'Z''$ and $P'''X''Y''Z''$; the perimeter of quadrilateral $PXYZ$ is clearly equal to the length of the polygonal line $PXY'Z''P'''$. From this it follows that if the line PP''' meets sides AB, BC' and $C'D''$ of the quadrilaterals that have been constructed, then the points of intersection determine the desired quadrilateral; if PP''' does not meet all three of these segments, then the desired quadrilateral will be degenerate (that is, a triangle, one of whose vertices coincides with one of the vertices of quadrilateral $ABCD$, or a diagonal, described twice, of quadrilateral $ABCD$).

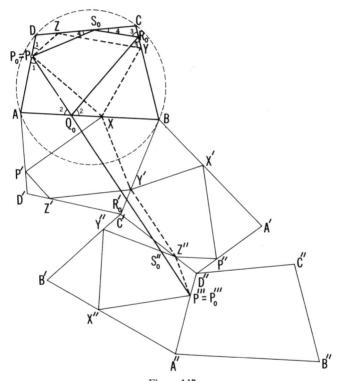

Figure 147a

Let us now pass to the solution of the original problem. We must find a point P on side AD of quadrilateral $ABCD$ such that the segment PP''', where P''' is the point corresponding to P on side $A''D''$ of quadrilateral $A''B'C'D''$, has minimum length [compare the solution to Problem 75 (b)]. As we saw in the solution to Problem 75 (b), if the segments AD and $A''D''$ are not parallel and do not have the same direction, then the desired point will be the foot of the perpendicular dropped from the rotation center of the segments AD and $A''D''$ onto the line AD (provided that this foot lies on the segment AD), or the desired point will be that endpoint of the segment AD that is closest to the foot of this perpendicular (if the foot of the perpendicular lies outside the segment AD). If the segments AD and $A''D''$ are parallel and have the same direction, then the distance from each point P of side AD to the corresponding point P' of segment $A''D''$ is the same.

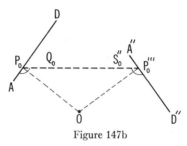

Figure 147b

The problem has a (unique) solution in the proper sense of the word only in the first case, if at the same time the foot P_0 of the perpendicular from the rotation center O of segments AD and $A''D''$ onto line AD lies on side AD, and if the line P_0P_0''' , where P_0'' is the foot of the perpendicular from O onto $A''D''$, meets the three segments AB, BC' and $C'D''$. In this case the points of intersection of the segment P_0P_0''' with the segments AB, BC', $C'D''$ define the desired quadrilateral $P_0Q_0R_0S_0$ (see Figure 147a). Moreover it is easy to see that $\sphericalangle P_0'''P_0A = \sphericalangle P_0P_0'''D''$ (see Figure 147b), from which it follows that sides P_0Q_0 and P_0S_0 of the desired quadrilateral form equal angles with side AD of the quadrilateral $ABCD$. Further, since the vertical angles that are formed by line P_0P_0''' with lines AB, BC' and $C'D''$ are equal, it follows that sides P_0Q_0 and Q_0R_0 of quadrilateral $P_0Q_0R_0S_0$ form equal angles with side AB of quadrilateral $ABCD$, that sides Q_0R_0 and R_0S_0 form equal angles with side BC, and that sides R_0S_0 and S_0P_0 form equal angles with side CD.† Further, with the notations of Figure 147a,

$$\sphericalangle A + \sphericalangle C = (180° - \sphericalangle 1 - \sphericalangle 2) + (180° - \sphericalangle 3 - \sphericalangle 4)$$
$$= 360° - (\sphericalangle 1 + \sphericalangle 2 + \sphericalangle 3 + \sphericalangle 4),$$
$$\sphericalangle B + \sphericalangle D = (180° - \sphericalangle 2 - \sphericalangle 3) + (180° - \sphericalangle 4 - \sphericalangle 1)$$
$$= 360° - (\sphericalangle 1 + \sphericalangle 2 + \sphericalangle 3 + \sphericalangle 4),$$
$$\sphericalangle A + \sphericalangle C = \sphericalangle B + \sphericalangle D = 180°.$$

† Thus we have shown that if $PQRS$ is a quadrilateral of *minimum perimeter* inscribed in $ABCD$, then any two consecutive sides of it form equal angles with that side of $ABCD$ which they meet. This proposition can be proved more simply. Indeed, if $\sphericalangle SPD \neq \sphericalangle QPA$, then, without changing the positions of vertices Q, R and S, we can change the position of P so that the perimeter of the quadrilateral $PQRS$ becomes smaller [see the solution to Problem 74 (a)].

Thus the problem has a solution in the proper sense of the word only if the quadrilateral $ABCD$ can be inscribed in a circle.

Now suppose that the quadrilateral $ABCD$ can be inscribed in a circle (that is, $\angle B + \angle D = 180°$); let $D''A''B''C''$ be the reflection of $D''A''B'C'$ in $D''A''$. Since quadrilateral $BC'D''A'$ is obtained from $ABCD$ by a rotation about the point B through an angle $2\angle B$ [compare the solution to Problem 75 (a)], and $D''A''B''C''$ is obtained from $D''A'BC'$ by a rotation about D'' through an angle $2\angle D$, and $2\angle B + 2\angle D = 360°$, it follows that $A''B''C''D''$ is obtained from $ABCD$ by a translation (see Section 2, Chapter I of Volume One, page 34). Therefore, if $ABCD$ can be inscribed in a circle then the segment $A''D''$ is parallel to AD and has the same direction. Thus the distance between an arbitrary point P of AD and the corresponding point P''' of $A''D''$ does not depend on the position of P, and thus our problem has an infinite set of solutions, corresponding to all those positions of P such that the segment PP''' meets the segments AB, BC' and $C'D''$. Clearly the sides of all of these quadrilaterals of minimum perimeter inscribed in $ABCD$ are parallel to one another (see Figure 148).

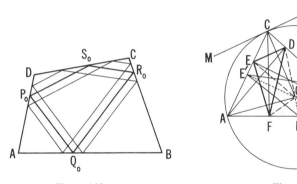

Figure 148   Figure 149

77. If D, E, F are the feet of the altitudes of $\triangle ABC$, then $\triangle ACD \sim \triangle BCE$ (see Figure 149); therefore $CE/CD = CB/CA$. Consequently, $\triangle ABC \sim \triangle DEC$ and $\angle CED = \angle CBA$. Let MN be the tangent to the circumscribed circle at the point C. Clearly $\angle MCA = \angle CBA$ ($=\frac{1}{2}$ arc AC). Thus, $\angle MCE = \angle CED$, that is, $ED \parallel MN \perp OC$. In the same way it can be shown that $EF \perp OA$, and $DF \perp OB$, that is, $\triangle DEF$ is the desired triangle. [It is easy to see that no other triangle inscribed in $\triangle ABC$ solves the problem, for no other inscribed triangle has its sides parallel to those of $\triangle DEF$.]

If $\triangle ABC$ is acute, then the center O of the circumscribed circle lies inside it (Figure 149); therefore

Area ($\triangle ABC$) = Area ($ODCE$) + Area ($OEAF$) + Area ($OFBD$).

But the diagonals in each of these last three quadrilaterals are perpendicular, and therefore their areas are equal to one half the product of their diagonals. Thus,

$$(*) \quad \text{Area} \, (\triangle ABC) = \tfrac{1}{2}OC \cdot DE + \tfrac{1}{2}OA \cdot EF + \tfrac{1}{2}OB \cdot FD$$
$$= \tfrac{1}{2}R(DE + EF + FD),$$

where R is the radius of the circumscribed circle.

If $F'D'E'$ is any inscribed triangle in ABC then

$$\text{Area} \, (\triangle ABC) = \text{Area} \, (OD'CE') + \text{Area} \, (OE'AF') + \text{Area} \, (OF'BD')$$
$$= \tfrac{1}{2}OC \cdot D'E' \sin \gamma + \tfrac{1}{2}OA \cdot E'F' \sin \alpha + \tfrac{1}{2}OB \cdot F'D' \sin \beta,$$

where γ, α and β are the angles between the diagonals of quadrilaterals $OD'CE'$, $OE'AF'$ and $OF'BD'$, respectively; therefore

$$\text{Area} \, (\triangle ABC) \le \tfrac{1}{2}R(D'E' + E'F' + F'D').$$

Comparing this equation with equation (*) above we see that

$$DE + EF + FD \le D'E' + E'F' + F'D'.$$

Thus, among all triangles inscribed in a given acute triangle, the one with the smallest perimeter is the one whose vertices are the feet of the altitudes of the given triangle.

78. We first find a point whose distances to the vertices A, B and C are proportional to the numbers a, b and c. The construction of such a point is easily accomplished if we use the fact that the locus of points the ratio of whose distances to two given points has a given value, is a circle (see the footnote on page 40). In addition it can be shown that in general there are two such points, one outside and one inside the circle circumscribed about the given triangle. Therefore one of these points certainly lies outside of $\triangle ABC$. We must now consider two cases.

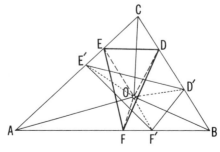

Figure 150

$1°$. One of the constructed points (we denote it by O) lies inside ABC (Figure 150). Inscribe in $\triangle ABC$ a triangle DEF whose sides are perpendicular to the segments OA, OB and OC [see Problem 9 (b), on page 17]. Clearly we have

$$\text{Area } (\triangle ABC) = \text{Area } (OEAF) + \text{Area } (OFBD) + \text{Area } (ODCE);$$

since $OA \perp EF$, $OB \perp FD$, $OC \perp DE$ and $OA = ak$, $OB = bk$, $OC = ck$ (for some number k), we have

$$(*) \qquad \text{Area } (\triangle ABC) = \tfrac{1}{2}OA \cdot EF + \tfrac{1}{2}OB \cdot FD + \tfrac{1}{2}OC \cdot DE$$
$$= \tfrac{1}{2}k(a \cdot EF + b \cdot FD + c \cdot DE).$$

Now let $D'E'F'$ be an arbitrary triangle inscribed in ABC, and let α, β, γ be the angles between the diagonals in quadrilaterals $OE'AF'$, $OF'BD'$, $OD'CE'$, respectively. In this case (compare the solution to Problem 77)

$$\text{Area } (\triangle ABC) = \tfrac{1}{2}OA \cdot E'F' \sin \alpha + \tfrac{1}{2}OB \cdot F'D' \sin \beta + \tfrac{1}{2}OC \cdot D'E' \sin \gamma$$
$$\leq \tfrac{1}{2}k(a \cdot E'F' + b \cdot F'D' + c \cdot D'E')$$

and therefore

$$a \cdot EF + b \cdot FD + c \cdot DE \leq a \cdot E'F' + b \cdot F'D' + c \cdot D'E',$$

that is, DEF is the desired triangle.

$2°$. If the point O lies, for example, on side AB of $\triangle ABC$, then the desired triangle degenerates to the altitude from C onto side AB, described twice; if O lies outside $\triangle ABC$, then the desired triangle degenerates to the altitude, described twice, on the side of $\triangle ABC$ that separates O from $\triangle ABC$†. We leave the proof to the reader.

79. The desired point M cannot lie outside of $\triangle ABC$, for otherwise we could easily find a point M' for which

$$AM' + BM' + CM' < AM + BM + CM$$

(Figure 151a). Now let X be an arbitrary point inside $\triangle ABC$ (Figure 151b). Rotate $\triangle ACX$ about A through an angle of $60°$ in the direction from B to C to the new position $AC'X'$. Since $AX = XX'$ (triangle AXX' is equilateral) and $CX = C'X'$, we see that the sum of the distances from the point X to the vertices of the triangle is equal to the

---

† Each line in the plane separates the plane into two half-planes; points which are in different half-planes are said to be separated from each other by the line. A point outside a given triangle is separated from the points inside the triangle by at least one, and at most two, of the sides (extended indefinitely) of the triangle. It can be shown that in our situation, the point O is separated from $\triangle ABC$ by only one side of $\triangle ABC$.

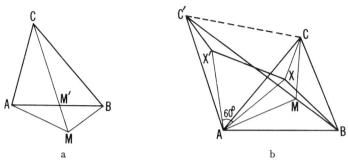

a   b

Figure 151

length of the polygonal line $BXX'C'$. It now remains to choose the point M so that the polygonal line $BMM'C'$ will have minimum length.

We consider separately two cases.

1°. *Segment $C'B$ intersects side AC of $\triangle ABC$*; this case obtains when $\sphericalangle BCC' < 180°$, $\sphericalangle BAC' < 180°$ or, since

$$\sphericalangle BCC' = \sphericalangle BCA + \sphericalangle ACC' = \sphericalangle BCA + 60°$$

and

$$\sphericalangle BAC' = \sphericalangle BAC + \sphericalangle CAC' = \sphericalangle BAC + 60°,$$

when angles C and A of the triangle are less than 120°. In this case if a point M can be found on the segment $C'B$ such that $\sphericalangle AMC' = 60°$, then for this point we will have

$$AM + MC + MB = C'B,$$

and, therefore, it will be the desired point. For this point M we have $\sphericalangle AMB = \sphericalangle CMB = \sphericalangle AMC = 120°$ (M is the point from which all sides of the triangle are visible under equal angles); clearly, for there to be such a point M on segment $C'B$ it is necessary that $\sphericalangle CBA$ be less than 120°. If angle $CBA \geq 120°$, then the desired point will be vertex B of triangle ABC.

2°. *Segment $C'B$ does not intersect side AC of $\triangle ABC$*; for example, the point C' lies on the opposite side of line BC from $\triangle ABC$ ($\sphericalangle C \geq 120°$). In thir case the shortest polygonal line $C'X'XB$ is the line $C'CB$, and the desired point is the vertex C of $\triangle ABC$. In the same way, if $\sphericalangle A \geq 120°$, then the desired point is vertex A.

80. (a) If DEF is an equilateral triangle circumscribed about $\triangle ABC$ and if AM, BM, CM are perpendicular to the sides of $\triangle DEF$ then, clearly, $\sphericalangle AMB = \sphericalangle BMC = \sphericalangle CMA = 120°$ (Figure 152a). It follows that M is the point of intersection of the circular arcs constructed on the sides of $\triangle ABC$ and each measuring an angle of 120°. Having found M we can construct $\triangle DEF$ without difficulty. The point M will lie inside $\triangle ABC$ if none of the angles of $\triangle ABC$ exceeds $120°$; if, for example, $\angle C = 120°$, then $M = C$; if $\angle C > 120°$, then M lies outside $\triangle ABC$. We now derive the solution to Problem 79 from this.

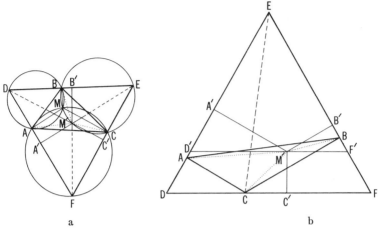

Figure 152

Consider the first case. Let M' be any interior point of $\triangle ABC$, and let $M'A'$, $M'B'$, $M'C'$ be the perpendiculars from M' onto the sides of $\triangle DEF$. We have

$$\text{Area} (\triangle DEF) = \text{Area} (\triangle DEM') + \text{Area} (\triangle EFM') + \text{Area} (\triangle FDM')$$

or, if a and h are the side and the altitude respectively of the equilateral triangle DEF,

$$\tfrac{1}{2}ah = \tfrac{1}{2}a \cdot M'A' + \tfrac{1}{2}a \cdot M'B' + \tfrac{1}{2}a \cdot M'C',$$

that is,

$$M'A' + M'B' + M'C' = h.$$

But $M'A \geq M'A'$, $M'B \geq M'B'$, $M'C \geq M'C'$ (since the perpendicular distance is shortest); therefore

$$M'A + M'B + M'C \geq h,$$

and the equality will hold only when M' coincides with M. Thus M is the desired point.

If $M = C$, then C is the desired point. Finally, if $\angle C > 120°$, then again the sum of the distances from the vertex C to the other vertices of the triangle will be less than the sum of the distances from any other point to the vertices of the triangle. To prove this circumscribe an isosceles triangle DEF about $\triangle ABC$ so that $CA \perp DE$, $CB \perp EF$ (Figure 152b). Let M' be an arbitrary interior point of $\triangle ABC$, and let A', B', C' be the feet of the perpendiculars from M' onto the sides of $\triangle DEF$; let

$D'EF'$ be the triangle similar to DEF whose base $D'F'$ passes through M'. Then we have

$$\text{Area } (\triangle DEF) = \text{Area } (\triangle CDE) + \text{Area } (\triangle CEF),$$

and so

$$CA + CB = h,$$

where h is the altitude of $\triangle DEF$ on one of its two equal sides. Similarly we obtain

$$M'A' + M'B' = h',$$

where $h' = kh$ is the altitude of triangle $D'EF'$ (k is the similarity coefficient, $k < 1$).

Denote the altitudes of triangles DEF and $D'EF'$ onto their respective bases by H and $H' = kH$. Since $\angle E = 180° - \angle C < 60°$, we have $DF < DE$, $H > h$. Therefore

$$M'A' + M'B' + M'C' = h' + (H - H') = H - (H' - h')$$

$$= H - k(H - h) > H - (H - h) = h = CA + CB.$$

Clearly we have

$$M'A' + M'B' + M'C' \leq M'A + M'B + M'C$$

and so

$$M'A + M'B + M'C > CA + CB,$$

which was to be proved.

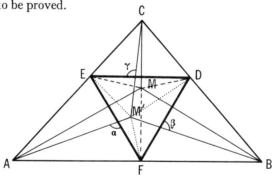

Figure 153

(b) If AM, BM and CM are perpendicular to the sides of the equilateral triangle DEF inscribed in $\triangle ABC$, then, clearly,

$$\angle AMB = \angle BMC = \angle CMA = 120°$$

(Figure 153). Therefore to solve the problem we must find a point M from which all sides of the triangle are visible under angles of $120°$, and then inscribe in $\triangle ABC$ a triangle DEF whose sides are perpendicular to AM, BM and CM [see Problem 9 (b), page 17]. At the same time, the vertices of $\triangle DEF$ will lie on the sides of $\triangle ABC$ and not on their extensions if all angles of $\triangle ABC$ are less than $120°$.

Assume that this is the case. Then

$$\text{Area } (\triangle ABC) = \text{Area } (MDCE) + \text{Area } (MEAF) + \text{Area } (MFBD)$$
$$= \tfrac{1}{2}DE \cdot MC + \tfrac{1}{2}EF \cdot MA + \tfrac{1}{2}FD \cdot MB$$
$$= \tfrac{1}{2}DE(MA + MB + MC).$$

Now let M' be an arbitrary point inside $\triangle ABC$ and let α, β, γ be the angles that the lines $M'A$, $M'B$ and $M'C$ make with the corresponding sides of $\triangle DEF$. Then

$$\text{Area } (\triangle ABC) = \text{Area } (M'DCE) + \text{Area } (M'EAF) + \text{Area } (M'FBD)$$
$$= \tfrac{1}{2}DE \cdot M'C \sin\gamma + \tfrac{1}{2}EF \cdot M'A \sin\alpha + \tfrac{1}{2}FD \cdot M'B \sin\beta$$
$$\leq \tfrac{1}{2}DE(M'A + M'B + M'C).$$

From this we have

$$MA + MB + MC \leq M'A + M'B + M'C$$

which was to be proved (compare the solution to Problem 77).

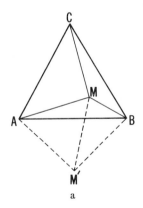
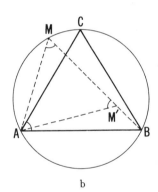

a
b

Figure 154

81. (a) *First solution.* Rotate $\triangle CAM$ about point A through $60°$ into the position ABM' (Figure 154a). Then $MM' = AM' = AM$, $CM = BM'$. But $BM \leq BM' + MM'$, and therefore

$$BM \leq AM + CM.$$

Moreover, $BM = BM' + MM'$ only if M' lies on the segment BM. Since $\measuredangle AMM' = 60°$, in this case we have $\measuredangle AMB = 60°$; this means that M lies on the arc AC of the circle circumscribed about triangle ABC (see Figure 154b).

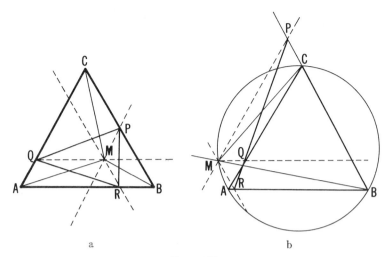

a   b

Figure 155

Second solution. Pass lines MP, MQ and MR through the point M parallel to sides AC, AB and BC of $\triangle ABC$ (Figure 155a). The quadrilaterals $MQAR$, $MPBR$, $MPCQ$ are easily seen to be isosceles trapezoids; consequently, $MA = QR$, $MB = PR$, $MC = PQ$. Thus the segments MA, MB and MC have the same lengths as the sides of $\triangle PQR$, and therefore $MA + MC \geq MB$.

The equality $MA + MC = MB$ holds only when $RQ + QP = PR$, that is, Q lies on the segment PR (Figure 155b). In this case $\angle RMA = \angle RQA$, $\angle PMC = \angle PQC$, $\angle RQA = \angle PQC$; that is, $\angle RMA = \angle CMP$, $\angle AMC = \angle RMP = 120°$ and, therefore, M lies on the arc AC of the circle circumscribed about $\triangle ABC$.

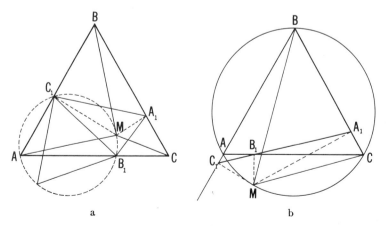

a   b

Figure 156

Third solution. From M drop perpendiculars MA_1, MB_1 and MC_1 onto the sides of $\triangle ABC$ (Figure 156a). The circle with diameter AM is circumscribed about the quadrilateral AC_1MB_1; since $\angle B_1AC_1 = 60°$, it follows that B_1C_1 is the side of an equilateral triangle inscribed in this circle. Therefore $B_1C_1 = (\sqrt{3}/2) \cdot MA$; similarly, $A_1B_1 = (\sqrt{3}/2) \cdot MC$, $A_1C_1 = (\sqrt{3}/2) \cdot MB$. But from triangle $A_1B_1C_1$ we have

$$A_1C_1 \leq A_1B_1 + B_1C_1,$$

from which

$$MB \leq MA + MC.$$

Moreover $MB = MA + MC$ if the points A_1, B_1 and C_1 lie on a line, with B_1 between A_1 and C_1 (Figure 156b). We assert that in this case M lies on the arc AC of the circle circumscribed about $\triangle ABC$. Indeed, $\angle C_1MA = \angle C_1B_1A$, $\angle A_1MC = \angle A_1B_1C$. But since in the case we are considering $\angle C_1B_1A = \angle A_1B_1C$, we have $\angle C_1MA = \angle A_1MC$ and $\angle AMC = \angle C_1MA_1 = 120°$, which proves our assertion.

We note that, in general, the feet of the perpendiculars from an arbitrary point M onto the sides of an arbitrary triangle are collinear if and only if M lies on the circle circumscribed about the triangle (see Problem 61 in Section 1 above).

Fourth solution. Let us apply Ptolemy's theorem (see Problem 269 of Section 4, Chapter II of Part Three of the Russian edition[T]) to quadrilateral $MABC$. We obtain

$$MB \cdot AC \leq MA \cdot CB + MC \cdot AB,$$

with equality holding if and only if a circle can be circumscribed about $MABC$. But $AC = CB = AB$ and, therefore,

$$MB \leq MA + MC,$$

which was to be proved.

(b) On side BC of the given triangle ABC, and outside of it, construct an equilateral triangle BCA' (Figure 157). Let X be an arbitrary point of the plane. From triangle XAA' we have

$$AA' \leq XA + XA',$$

and equality holds only if X lies on the segment AA'. Further, by the proposition of part (a),

$$XA' \leq XB + XC,$$

and equality holds only if X lies on the arc CmB of the circle circumscribed about triangle BCA'.

[T] This theorem states that for any quadrilateral with consecutive vertices A, B, C, D the sum of the products of the two pairs of opposite sides is greater than or equal to the product of the diagonals, that is,

$$AC \cdot BD \leq AB \cdot CD + AD \cdot BC,$$

with equality holding if and only if a circle can be circumscribed about the quadrilateral.

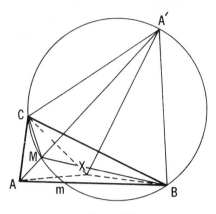

Figure 157

From these relations we have

$$AA' \leq XA + XB + XC.$$

If M is the point of intersection of AA' with the circumscribed circle of $\triangle A'BC$, then

$$AA' = MA + MB + MC,$$

that is

$$MA + MB + MC \leq XA + XB + XC,$$

and therefore M satisfies the conditions of Problem 79. By an easy calculation one can convince oneself that if one of the angles of $\triangle ABC$ is equal to $120°$, then the segment AA' and the arc CmB intersect at the vertex of this angle, while if one of the angles of the triangle is greater than $120°$, then they do not intersect at all. In this latter case it can be shown that the vertex of the obtuse angle is still the solution to the minimum problem.

82. First note that the desired point M must lie outside of $\triangle ABC$ and inside of the angle ACB. Indeed, suppose that M were an interior point of $\triangle ABC$, and let M' be the point of intersection of the line CM with side AB (Figure 158a). Then $AM' + BM' < AM + BM$ and $CM' > CM$, so that

$$AM' + BM' - CM' < AM + BM - CM.$$

Next assume that M does not lie inside $\sphericalangle ACB$. Then there are several possibilities for M. First assume that M lies in the angle that is vertical with respect to angle ACB, and let M' be the point symmetric to M with respect to the line l through C and parallel to AB (Figure 158b).

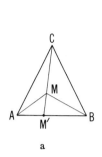

a

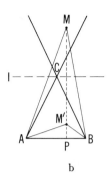

b

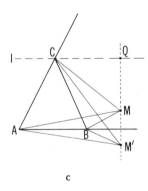

c

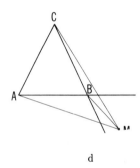

d

Figure 158

Then $M'C = MC$ and $M'A < MA$, $M'B < MB$ (these last two inequalities follow from the fact that, with the notations of Figure 158b, $M'P < MP$), and therefore

$$AM' + BM' - CM' < AM + BM - CM.$$

Next assume that M belongs to angle CAB (but does not lie on the line AB!) and does not lie inside triangle ABC, and let M' be the point symmetric to M with respect to line AB (Figure 158c). Then $AM' = AM$, $BM' = BM$, and $CM' > CM$ (the last inequality follows from the fact that, with the notations of Figure 158c, $M'Q > MQ$), and therefore

$$AM' + BM' - CM' < AM + BM - CM.$$

Similarly, the assumption that M belongs to $\sphericalangle CBA$ but not to $\triangle ABC$ leads to a contradiction. Finally, assume that M lies in the angle that is vertical with respect to $\sphericalangle ABC$ (or on the line AB). Then $MC - MB < BC$

and $MA > BA$ [the last inequality follows from the fact that $\angle MBA > \angle DBA > 90°$ (Figure 158d)], and therefore

$$MA + MB - MC = MA - (MC - MB)$$
$$> BA - BC = BA + BB - BC.$$

In the same way it can be shown that M cannot lie in the angle that is vertical with respect to angle BAC. Thus the assumption that M does not lie inside angle ACB has been contradicted.

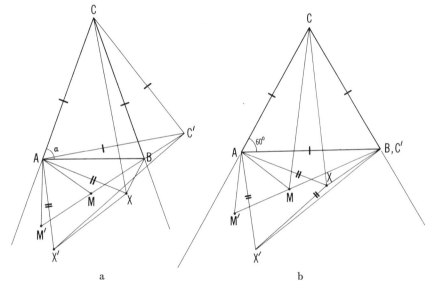

a   b

Figure 159

Now let X be an arbitrary point of the angle ACB that does not belong to $\triangle ABC$. Rotate $\triangle ACX$ about A through an angle of 60° in the direction from C to B, into position $AC'X'$ (Figure 159a). Since $AX = XX'$ (because triangle AXX' is equilateral) and $CX = C'X'$, it follows that $AX + BX - CX$ is equal to $X'X + BX - C'X'$; thus we must choose the point X so that the quantity $BX + XX' - C'X'$ is as small as possible. But since clearly

$$C'B + BX + XX' \geq C'X',$$

we therefore always have

$$BX + XX' - C'X' \geq -C'B;$$

thus if we can find a point M such that

(*) $C'B + BM + MM' = C'M'$ and $BM + MM' - C'M' = -C'B$,

where M' is obtained from M in the same way that X' was obtained from X, then M will be the desired point.

Now it is necessary to consider two cases.

1°. $\angle A = \alpha > 60°$, that is, $AC = BC > AB$ and $\triangle ABC$ is not equilateral. In this case C' *does not coincide* with B, and the equations (*) will hold provided that the points M and M' both lie on the line $C'B$ (Figure 159a). Since angle C of triangle ABC is equal to $180° - 2\alpha$, it follows that the angle at vertex C in triangle BCC' is equal to $60° - (180° - 2\alpha) = 2\alpha - 120°$; therefore

$$\angle CC'B = \angle CBC' = 150° - \alpha,$$

and so $\angle C'BA = \angle C'BC + \alpha = 150°$. Thus if M is chosen so that $\angle C'BM = 180°$, then we will have $\angle ABM = 30°$. If in addition M is chosen so that $\angle BMM' = \angle BMA + 60° = 180°$, then $\angle BMA = 120°$. From this it follows that there is a unique point M such that M and M' both lie on the line $C'B$ and

$$AM + BM - CM = BM + MM' - C'M = -C'B;$$

this point is characterized by the condition $\angle MBA = \angle MAB = 30°$ (Figure 160a).

2°. $\angle A = 60°$, that is, $\triangle ABC$ is equilateral; in this case $C' = B$ (Figure 159b). Also we have $BM + MM' - C'M' = -C'B$ $(=0)$, if the broken line BMM' is actually a segment of a line, and therefore $\angle BMA = 120°$ (since $\angle AMM' = 60°$). All such points M (see Figure 160b) lie on the arc AB of the circle circumscribed about $\triangle ABC$; each such point satisfies the requirements of the problem [compare the solution to Problem 81 (a)].

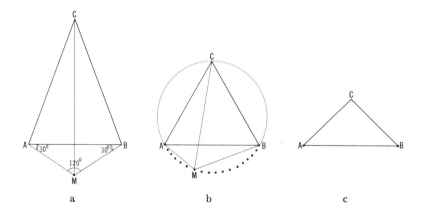

Figure 160

Remark. The solution given above does not work in case $AC = BC < AB$ and $\angle A = \alpha < 60°$. In this case it can be shown that the minimum of the expression $AM + BM - CM$ is attained when the point M coincides with vertex A or with vertex B of $\triangle ABC$ (Figure 160c).

One could also pose the problem of finding a point M for which the expression $MA + MB - MC$ is as small as possible, when $\triangle ABC$ is an *arbitrary* triangle given in advance. Here too if equation (*) is valid for some point M, lying in angle ACB but not in $\triangle ABC$, then it is not difficult to see that if M also satisfies

$$\angle AMC = \angle BMC = 60°$$

and therefore $\angle AMB = 120°$, then for this point the expression

$$MA + MB - MC(= -C'B)$$

will be as small as possible. [Here the point C' is obtained from C by a rotation about A through 60° in the direction from AC to AB; this rotation carries M into the point M' that occurs in the equations (*).] However, a complete description of the conditions when equation (*) is possible is quite difficult in general.

83. First consider the case when the sum of the two smaller of the numbers a, b, c does not exceed the third; suppose, for example, that $a \geq b + c$. For any point X we have

$$a \cdot XA + b \cdot XB + c \cdot XC \geq (b + c)XA + b \cdot XB + c \cdot XC$$

$$= b(XA + XB) + c(XA + XC) \geq b \cdot AB + c \cdot AC$$

(because $XA + XB \geq AB$, $XA + XC \geq AC$), so that the sum $a \cdot XA + b \cdot XB + c \cdot XC$ takes the smallest possible value when the point X coincides with the point A.

Thus it only remains to consider the case when there exists a triangle with sides equal to a, b and c. In considering this case we may follow four paths, analogous to the solutions to Problems 79, 80 (a), 80 (b) and 81.

First solution. Let $A_0B_0C_0$ be a triangle with sides equal to a, b and c, and let $\alpha = a/b$, $\gamma = c/b$; let X be an arbitrary point in the plane. The spiral similarity with center A, similarity coefficient γ and rotation angle equal to angle A_0 of $\triangle A_0B_0C_0$ (the rotation is taken in the direction from B to C), carries triangle AXC into $\triangle AX'C'$ (Figure 161). Triangles $AX'X$ and $A_0B_0C_0$ are similar since, by hypothesis,

$$AX'/AX = \gamma = A_0B_0/A_0C_0, \qquad \angle XAX' = \angle B_0A_0C_0.$$

From their similarity we have $XX'/AX = a/b = \alpha$; $XX' = \alpha AX$; further, by construction, $C'X' = \gamma CX$. Consequently,

$$C'X' + X'X + XB = \gamma \cdot CX + \alpha \cdot AX + BX = \frac{c \cdot CX + a \cdot AX + b \cdot BX}{b}$$

and, therefore, the quantity $a \cdot AX + b \cdot BX + c \cdot CX$ has the smallest value when the polygonal line $BXX'C'$ has the shortest length. The following cases are possible here.

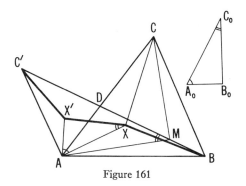

Figure 161

1°. Line BC' meets side AC of the given triangle in some point D. In this case the shortest polygonal line joining the points B and C' and crossing the segment AC, is the segment BC'. Using the fact that angle AXX' is equal to angle C_0 of $\triangle A_0 B_0 C_0$, it is easy to find the point M. To do this we construct an arc on the segment AD that measures the indicated angle and that lies on the same side of line AC as the point B. If this arc intersects the segment BC' then the point of intersection is the desired point M. If the arc does not intersect the segment BC' then the desired point M coincides with B.

2°. If line BC' does not meet side AC of $\triangle ABC$, then the shortest polygonal line $BXX'C'$ that meets side AC will be either the polygonal line BCC' or the polygonal line BAC'. Clearly in the first case $M = C$, and in the second case $M = A$.

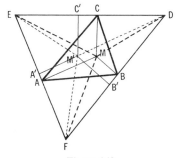

Figure 162

Second solution. If in $\triangle DEF$ we have $EF:FD:DE = a:b:c$, then the sum of the distances to the sides of $\triangle DEF$ from an arbitrary point M, multiplied respectively by the numbers a, b and c, is constant. Indeed, from Figure 162 we have

Area $(\triangle DEF)$ = Area $(\triangle MEF)$ + Area $(\triangle MFD)$ + Area $(\triangle MDE)$

or, if $EF = ak$, $FD = bk$, $DE = ck$,

$$\text{Area } (\triangle DEF) = \tfrac{1}{2}MA \cdot ka + \tfrac{1}{2}MB \cdot kb + \tfrac{1}{2}MC \cdot kc,$$

$$a \cdot MA + b \cdot BM + c \cdot MC = \frac{2 \text{ Area } (\triangle DEF)}{k} = \text{constant.}$$

Here A, B and C are the feet of the perpendiculars from M to the sides of triangle DEF.

Now circumscribe about $\triangle ABC$ a triangle DEF whose sides are in the ratio $a:b:c$, such that the perpendiculars to the sides of DEF, erected at the points A, B and C, meet in a common point M [the construction is similar to the construction in Problem 80 (a)]. If M lies inside $\triangle ABC$ then it is the solution to our minimum problem; the proof of this proceeds as in the solution to Problem 80 (a). If M lies outside $\triangle ABC$, then the solution to the problem is given by one of the vertices of the triangle.

Third solution. Inscribe in the given triangle ABC a triangle DEF with sides in the ratios $a:b:c$ and with the property that the perpendiculars dropped from the vertices of $\triangle ABC$ to the sides of $\triangle DEF$ meet in a common point M [the construction is similar to the construction in Exercise 80 (b)].

In case $\triangle DEF$ is inscribed in $\triangle ABC$ in the ordinary sense (that is, all its vertices lie on the sides of $\triangle ABC$ and not on their extensions), then just as in the solution to Problem 80 (b) it can be shown that M is the desired point. Otherwise the solution to the problem is given by one of the vertices of $\triangle ABC$.

Fourth solution. We make use of the following proposition, which generalizes the result of Problem 81 (a): if in $\triangle ABC$ we have $BC:CA:AB = a:b:c$, then for any point M in the plane we have

$$b \cdot MB \leq a \cdot MA + c \cdot MC,$$

and equality holds when M lies on the corresponding arc of the circumscribed circle of $\triangle ABC$. The proof of this proposition [which can be carried out in several ways, analogous to the solutions to Problem 81 (a)] we leave to the reader.

We now construct on side BC of the given triangle a triangle BCA' such that $BC:CA':A'B = a:b:c$, and we circumscribe a circle about this triangle. If X is an arbitrary point in the plane, then from $\triangle XAA'$ it follows that

$$AA' \leq XA + XA',$$

and equality holds only for points on the segment AA'. In addition,

$$a \cdot XA' \leq b \cdot XB + c \cdot XC,$$

and equality holds only for points of the arc BmC.

Multiplying the first of these relations by a and adding it to the second we obtain

$$a \cdot AA' \leq a \cdot XA + b \cdot XB + c \cdot XC,$$

and equality holds only for the point M of intersection of the arc BmC with the segment AA':

$$a \cdot AA' = a \cdot MA + b \cdot MB + c \cdot MC.$$

Thus,

$$a \cdot MA + b \cdot MB + c \cdot MC \leq a \cdot XA + b \cdot XB + c \cdot XC,$$

that is, M is the solution to the problem.

If the arc BmC does not intersect the segment AA', then it can be shown that the solution to the problem is given by one of the vertices of $\triangle ABC$.